T0224668

CONCEPTIONS OF SET AND THE FOUNDATIONS OF MATHEMATICS

Sets are central to mathematics and its foundations, but what are they? In this book Luca Incurvati provides a detailed examination of all the major conceptions of set and discusses their virtues and shortcomings, as well as introducing the fundamentals of the alternative set theories with which these conceptions are associated. He shows that the conceptual landscape includes not only the naïve and iterative conceptions but also the limitation of size conception, the definite conception, the stratified conception and the graph conception. In addition, he presents a novel, minimalist account of the iterative conception which does not require the existence of a relation of metaphysical dependence between a set and its members. His book will be of interest to researchers and advanced students in logic and the philosophy of mathematics.

LUCA INCURVATI is Assistant Professor of Philosophy at the University of Amsterdam.

CONCEPTIONS OF SET AND THE FOUNDATIONS OF MATHEMATICS

LUCA INCURVATI

University of Amsterdam

CAMBRIDGE
UNIVERSITY PRESS

CAMBRIDGE
UNIVERSITY PRESS

University Printing House, Cambridge CB2 8BS, United Kingdom

One Liberty Plaza, 20th Floor, New York, NY 10006, USA

477 Williamstown Road, Port Melbourne, VIC 3207, Australia

314-321, 3rd Floor, Plot 3, Splendor Forum, Jasola District Centre, New Delhi - 110025, India

79 Anson Road, #06-04/06, Singapore 079906

Cambridge University Press is part of the University of Cambridge.

It furthers the University's mission by disseminating knowledge in the pursuit of
education, learning and research at the highest international levels of excellence.

www.cambridge.org
Information on this title: www.cambridge.org/9781108708791
DOI: 10.1017/9781108596961

First published 2020
First paperback edition 2021

A catalogue record for this publication is available from the British Library

ISBN 978-1-108-49782-4 Hardback
ISBN 978-1-108-70879-1 Paperback

Contents

Figures

Tables

Preface

I took my first set theory course as a philosophy student at the Department of Mathematics of the University of Rome La Sapienza. It was after taking that course, taught by Professor Claudio Bernardi, that I decided to dedicate more time and energy to the philosophy of set theory if I could.

The course was entitled 'Foundations of Mathematics' and had one striking feature: it was taught from the ZFC axioms, with almost no mention of naïve set theory and the set-theoretic paradoxes. When I went up to Cambridge in 2005, I began to read some of the standard textbooks on set theory, such as Devlin 1993 and Jech 2003. Naïve set theory was now introduced and the paradoxes were given their due. But after this, ZFC was presented, and one was left with the impression that all roads from naïve set theory lead to the cumulative hierarchy.

This book is an attempt to reverse this trend. Perhaps, in the end, we should stick with the iterative conception of set. Indeed, one can read the book as an extended argument to this effect. But the path from the paradoxes to ZFC and cognate systems is much more tortuous than tradition has made it out to be.

The book describes and assesses a number of *conceptions of set*. Being a book on the *philosophy* of set theory, the focus is very much on the philosophical underpinnings of these conceptions. But history is important, in philosophy as in life, and I have provided some historical background when I deemed it useful.

Being a book on the philosophy of *set theory*, the book deals with technical material. I have tried to make the material as accessible as possible and have included some sections introducing some of the basic notions used, namely Section 1.1 and the appendices to Chapter 1. More advanced notions from set theory are introduced in the body of the text or in footnotes.

The book deals with some theories that are likely to be less known to philosophical audiences (e.g. the non-well-founded set theories of Chapter 7) and mathematical audiences (e.g. certain set theories based on the naïve conception of set, presented in Chapter 4). The book can also serve as a friendly introduction to these set theories, whose central formal aspects are described in the relevant chapters.

Parts of the book are based on material which has previously appeared in print. In particular, Chapters 2 and 7 draw, respectively, on 'How to be a minimalist about sets' (*Philosophical Studies*, 2012) and 'The graph conception of set' (*Journal of Philosophical Logic*, 2014). I am grateful to Springer for permission to use material from these two articles. In addition, parts of Section 3.6 were originally published as section 2 of 'Maximality principles in set theory' (*Philosophia Mathematica*, 2017), and I would like to thank Oxford University Press for permission here.

* * *

A summary of the content of the individual chapters will be helpful. The first chapter explains what conceptions of set are, what they do, and what they are for. It begins by introducing the concept of set and making explicit certain assumptions about concepts. Following a tradition going back to Frege, concepts are taken to be equipped with criteria of application and, in some cases, criteria of identity. It is pointed out that a concept need not settle all questions concerning which objects it applies to and under what conditions those objects are identical. This observation is used to characterize conceptions: a conception settles more questions of this kind than the corresponding concept does. On this account, a conception *sharpens* the corresponding concept. This view is contrasted with two other possible views on what conceptions do: one according to which they provide an *analysis* of what belongs to a concept, and one according to which they are possible *replacements* for a given concept. The chapter concludes with a discussion of the uses of conceptions of set. Along the way, the *naïve conception of set*, which holds that every condition determines a set, is introduced and a diagnosis of the set-theoretic paradoxes is offered. According to this diagnosis, the naïve conception leads to paradox because it requires the concept of set to have two features – *indefinite extensibility* and *universality* – which are jointly inconsistent.

The second chapter introduces the *iterative conception*, according to which every set appears at one level or another of the mathematical structure known as the *cumulative hierarchy*, as well as theories based on the conception. The chapter presents various accounts of the iterative conception: the *constructivist* account, the *dependency* account and my own *minimalist* account. It is argued that the minimalist account is to be preferred to the others. A method – which I call *inference to the best conception* – is then described to defend the correctness of the iterative conception so understood. This method requires one to show that the iterative conception fares better than other conceptions with respect to a number of desiderata on conceptions of set. This provides additional motivation for exploring alternative conceptions of set in the remainder of the book.

The main purpose of the third chapter is to defend the iterative conception against three objections. The first objection, which I call the *missing explanation objection*, is that if the iterative conception is correct, one cannot explain the intuitive appeal of the Naïve Comprehension Schema. The chapter provides plausible explanations of this fact which are compatible with the correctness of the conception. The second objection, which I call the *circularity objection*, is that the iterative conception presupposes the notion of an ordinal, and since ordinals are treated in set theory like certain kinds of sets, this means that the conception is vitiated by circularity. The chapter shows that this objection can be defeated by constructing ordinals using a trick that goes back to Tarski and Scott or dispensing with the notion of well-ordering altogether in the formulation of the conception. The third objection, which I call the *no semantics objection*, is that the iterative conception prevents us from giving a semantics for set theory, since according to the conception there is no universal set but we seem to need such a set to serve as the domain of quantification when giving a semantics for set theory. The chapter defends the approach that this problem can be overcome by doing semantics in a higher-order language. The chapter concludes by discussing the status of the Axiom of Replacement on the iterative conception.

The fourth chapter discusses the naïve conception of set and criticizes attempts to rehabilitate it by modifying the logic of set theory. The focus is on the proposal that the Naïve Comprehension Schema – which formally captures the thesis that every condition determines a set – is to be saved by adopting a paraconsistent logic. Three strategies for doing so are distinguished: the *material* strategy, the *relevant* strategy and the *model-theoretic strategy*. It is shown that these strategies lead to set theories that are either too weak or ad hoc or give up on the idea that sets are genuinely extensional entities.

Chapters 5 and 6 consider attempts to modify the naïve conception of set by restricting it in an appropriate manner. The idea is that the core thought of the naïve conception – that there is an intimate connection between sets and properties – can be preserved as long as we build into the conception the idea that certain properties are pathological and, for this reason, do not determine a set. The fifth chapter first uses a result from Incurvati and Murzi 2017 to show that restricting attention to those properties that do not give rise to inconsistency will not do. It then focuses on the *limitation of size conception of set*, according to which the pathological properties are those that apply to too many things. Various versions of the doctrine are distinguished. The chapter also discusses what it calls the *definite conception*, according to which the pathological properties are the indefinitely extensible ones. It is argued that the limitation of size fails to provide a complete explanation of the set-theoretic paradoxes. The definite conception faces the same problem and, in addition, it is unclear whether it has the resources to develop a reasonable amount of set theory.

The sixth chapter considers attempts to modify the Naïve Comprehension Schema according to syntactic notions and in particular the technical notion of stratification used in Quine's New Foundations. This and related theories (such as New Foundations with *Urelemente*) are examined and their philosophical underpinnings discussed and assessed. It is argued that, contra what has often been suggested in the literature, one can describe a conception of set – the *stratified conception* – which is well motivated whilst incorporating the idea that the pathological properties are those whose syntactic expression does not satisfy the stratification requirement. However, it is argued that this conception is best seen as a conception of objectified properties rather than as a conception of sets as combinatorial collections. Understanding New Foundations and cognate systems as theories of objectified properties leads to a further development of the diagnosis of the set-theoretic paradoxes offered in the first chapter and allows one to deal with some of the standard objections to set theories based on stratification.

The seventh chapter presents and discusses the *graph conception of set*. According-ing to the graph conception, sets are things depicted by graphs of a certain sort. The chapter begins by presenting four set theories, due to Aczel, which are for-mulated by using the notion of a graph. The graph conception is then introduced, and a historical excursion into forerunners of the conception is also given. The chapter continues by clarifying the relationship between the conception and the four theories described by Aczel. It concludes by discussing four objections to the graph conception: the objection that set theories based on graphs do not introduce new isomorphism types; the objection that the graph conception does not provide us with an intuitive model for the set theory it sanctions; the objection that the graph conception cannot naturally allow for *Urelemente*; and the objection that a set theory based on the graph conception cannot provide an autonomous foundation for mathematics. It is argued that whilst the first two objections fail, the remaining two retain their force.

The eighth and last chapter offers some concluding remarks. It provides an overview of some of the central features of the conceptions of set encountered in the book. It then discusses to what extent the findings of the book are compatible with some form of pluralism about conceptions.

* * *

This book is the result of thinking that spans more than a decade. As my ideas evolved, I was fortunate enough to be able to present them to a number of audi-ences who provided useful feedback. These include the Moral Sciences Club in

Cambridge, the Foundations of Mathematics Seminar at the University of Paris 7-Diderot, the seminar of the project *Plurals, Predicates and Paradox* in London, the Third Paris-Nancy Philosophy of Mathematics Workshop, the Philosophy of Mathematics Seminar in Oxford, the Faculty of Philosophy of the Università Vita-Salute San Raffaele, the 88th Joint Session of the Aristotelian Society and the Mind Association, the Department of Philosophy of the University of Amsterdam and the seminar of the Logic Group of the University of Hamburg.

The following set of *Urelemente* deserves special mention: {Arianna Betti, Robert Black, Francesca Boccuni, Tim Button, Mic Detlefsen, Salvatore Florio, Volker Halbach, Leon Horsten, Dan Isaacson, Hannes Leitgeb, Øystein Linnebo, Benedikt Löwe, Julien Murzi, Alex Oliver, Alexander Paseau, Michael Potter, Ian Rumfitt, Stewart Shapiro, Rob Trueman, Sean Walsh, Nathan Wildman, Tim Williamson}. No doubt this set is a proper subset of the set of *Urelemente* to whom I am indebted. I am also grateful to Hilary Gaskin at Cambridge University Press for her help and support throughout this project and to two referees, whose detailed and insightful comments on earlier versions of this material led to its final shape. Finally, I would like to thank my parents and Sarah for their constant and loving support. This book is dedicated to the memory of my late grandparents.

1

Concepts and Conceptions

Class *presented itself as a vague notion,*
or, specifically, a mixture of notions.

Georg Kreisel (1967: 82)

This book, as its title indicates, is about *conceptions of set*. Conceptions of set are advocated by mathematicians and philosophers with different, but often related purposes in mind. In this chapter, we offer a preliminary discussion of what we can take conceptions of set to be, what they are supposed to do, and the goals they are supposed to achieve. In so doing, we will provide a map of the wider territory within which the debate about conceptions of set takes place. Along the way, we shall briefly encounter our first conception of set, namely the *naïve conception*.

1.1 Theories

Before we get started, however, it will be helpful to remind the reader of a few notions we will use throughout. This will also help to fix terminology.

Conceptions of set, as we shall see, are often invoked to justify *theories* of set. At the most general level, a *theory* T in a given language consists of some sentences of that language. But the theories that conceptions of set are invoked to justify are typically *axiomatic formal theories*.

Axiomatic formal theories are *axiomatic* because they have (non-logical) *axioms*. That is to say, some of their sentences are singled out as those sentences which are not in need of proof and from which the theorems are derived. These derivations are carried out using a particular *logic*. For the most part, we will be concerned with *classical first-order logic*, but we will have occasion to look at theories that are cast in logics which extend this logic or deviate from it. For instance, we will look at theories that are cast in classical *second-order logic*, which we shall distinguish by adding a '2' as a subscript to their name, and at theories that are cast in *paraconsistent* logic. Relevant details will be given in due course.

Axiomatic formal theories are *formal* because the language they are cast in is a *formalized* language. A formalized language $\mathcal{L}$ is such that there are precise rules for determining which sequences of symbols constitute a *well-formed formula* of $\mathcal{L}$, and a sentence of $\mathcal{L}$ is simply a well-formed formula with no free variables. The symbols of a formalized language can be divided into those that make up its *logical vocabulary* – in our case, variables, bracketing devices and symbols for connectives, quantifiers and identity – and those that make up its *non-logical vocabulary* – for instance, individual constants and predicates.

Sometimes, when talking of a formalized language one is only referring to the syntax of that language. Other times, however, one is referring to the language as endowed with meaning, and, in the context of axiomatic formal theories, we will be using the term 'language' in both ways. In the latter case, the idea is that the language has a particular *interpretation*. Standardly, an *interpretation* for a given language $\mathcal{L}$ is an ordered pair $\mathfrak{A} = \langle \mathcal{D}, \mathcal{I} \rangle$ where $\mathcal{D}$ is the domain of interpretation – usually thought of as a non-empty set – and $\mathcal{I}$ is an interpretation function, i.e. a function satisfying the following conditions:

- $\mathcal{I}(c)$ is an element of $\mathcal{D}$ when c is a name of $\mathcal{L}$;
- $\mathcal{I}(E)$ is a set of n-tuples of $\mathcal{D}$-elements when E is an n-place predicate.

A *model* of a theory is then an interpretation which makes every sentence of the theory true.[1] Note that this means that, officially, an interpretation is a set-theoretic entity. In Section 3.4, we will discuss an approach to model theory which is different in this respect.

Finally, we will follow custom and will sometimes use T to refer solely to the axioms of T rather than to the deductive closure of these axioms – that is, the theorems that can be derived from T's axioms in T's logic. The context will make it clear when this is the case.

1.2 The Concept of Set

Unsurprisingly, conceptions of set are best explained by reference to the *concept* of set. On the face of it, English speakers possess this concept: if I say that the set of books on my table has two elements, you understand what I am saying. Moreover, English speakers with some knowledge of basic, secondary school mathematics

---

[1] For more on the notions of model and truth in an interpretation, see Section 3.3.

seem to understand sentences such as 'the set of natural numbers is infinite' and 'every set of real numbers which has an upper bound has a least upper bound'. Thus, not only do speakers seem to possess the concept of set as it occurs in everyday parlance, but they also seem to have the concept of set as it occurs in elementary mathematics.

Things, however, are not as simple as they may seem at first sight. Quite often, when people talk of a set of things, what they say can easily be recast without any mention of sets. So, for instance, when I said that the set of books on my table has two elements, it would have been prima facie legitimate for someone to take me simply to be saying that there are two books on my table. In other words, 'set of things' is often used, in ordinary parlance, as synonymous for what philosophers call a *plurality* of these things. A plurality, in this sense, is not something over and above the things comprising it: it is just them.

Other terms used by philosophers and mathematicians to talk about the same sort of entities are 'totality', 'multiplicity', 'multitude' and, sometimes, 'class'. With regard to the latter, Paul Finsler, as early as 1926, makes the point as follows:

It would surely be inconvenient if one always had to speak of many things in the plural; it is much more convenient to use the singular and speak of them as a *class*. [...] A class of things is understood as being the things themselves, while the set which contains them as its elements is a single thing, in general distinct from the things comprising it. [...] Thus a set is a genuine, individual entity. By contrast, a class is singular only by virtue of linguistic usage; in actuality, it almost always signifies a plurality. (Finsler 1926: 106)

Depending on what stance one takes on whether there are pluralities consisting of just one object or of no object at all, pluralities may be regarded as either a special case of, or identified with, what Bertrand Russell (1903) calls *classes as many*. The notion of a class as many is best understood by distinguishing between three cases in which we seem to have the class as many of the things falling under a concept C.

First, there is the case in which nothing falls under C. In this case, there is no class as many of the things falling under C. As Russell (1903: §69) puts it, 'there is no such thing as the null class, though there are null class-concepts'. Second, there is the case in which exactly one thing falls under C. Letting $[x: x$ is $\Phi]$ denote the concept under which the things that are Φ fall (or the property had by the Φ-things), an example is provided by the concept $[x: x$ is a current Chancellor of Germany]. In this case, the class as many of the things falling under C is the thing itself – Angela Merkel, in the case we are considering (as of 2019). Using Russell's (1903: §69) words: 'a class having only one term is to be identified [...] with that one term'. Third, there is the case in which more than one objects falls under the concept whose class as many we are considering. In that case, the class as many is just the plurality of the things falling under that concept. And, as Russell observes, '[i]n such cases, though terms may be said to belong to the class, the class must not be treated as itself a single logical subject' (Russell 1903: §70).

In general, then, a class as many is to be identified with the things falling under a certain concept, be they none, one or many. Bearing this fact in mind, we can now begin to say a bit more about the concept we are focusing on in this book, the concept of set. For one central feature of sets is that they are single objects, they are *unities*, even in the case in which there is more than one object comprising them. A set is a 'genuine, individual entity', as Finsler put it in the quote above – it is a *class as one*, to use Russell's terminology (1903: §74). Thus, in at least certain cases, some things yy will form a set, which is a single entity. Such a set will be the set a of all x such that x is one of the yy, which we write $\{x \mid x$ is one of the $yy\}$. We say that each b among the yy is a *member* of (or *element* of) a, which we write $b \in a$. Similarly, we write $b \notin a$ to indicate that b is *not* a member of a.

Let us now return to the example of the set of the natural numbers. The idea is that the plurality of my hands, which for the occasion we may call Left and Right, just is my hands – just is Left and Right. The set {Left, Right}, on the other hand, is a single object, whose members are Left and Right. Or consider again our earlier example about the set of natural numbers. Although when we talk about this set there is a reading according to which we are talking about *the natural numbers themselves*, we will be concerned with the reading according to which we are talking about *the set*, understood as a genuine entity over and above the natural numbers.

This central aspect of the concept of set was emphasized time and again by the founder of set theory, Georg Cantor. In an oft-quoted passage he writes:

By a 'set' we understand every collection to a whole M of definite, well-differentiated objects m of our intuition or our thought. (Cantor 1895: 282)

Similar remarks appear elsewhere. For instance, two years earlier Cantor had written:

By a 'manifold' or 'set' I understand in general any many [*Viele*] which can be thought of as one [*Eines*], that is, every totality of definite elements which can be united to a whole through a law. (Cantor 1883: 204, fn. 1)

And in a 1899 letter to Dedekind he tells us that

[i]f [...] the totality of the elements of a multiplicity can be thought of without contradiction as 'being together', so that they can be gathered together into '*one* thing', I call it a *consistent multiplicity* or a 'set'. (Cantor 1899: 114)

The last characterization of the notion of set appeals to consistency, and we shall return to this aspect of Cantor's thought in Chapter 5. The second quotation appeals to the idea that it is through a *law* that a plurality becomes a unity – an idea which is absent in the other quotes. Finally, in all passages some mention is made of the role that our *thought* has in determining whether a certain plurality forms a set, which is something that – as we shall see in the next chapter – raises a number of issues. But the feature of sets which is mentioned in all passages and which, following Cantor, is now standardly taken to be part of the concept of set and distinguishes from the concept of a plurality is that a set is *one thing*.

Cantor is also explicit that, as we emphasized, a set is an entity over and above its members:

This set [the set of all natural numbers] is a thing in itself and constitutes, completely apart from the natural sequence of numbers belonging to it, a firm in all its parts, determinate quantum, an *apōrisménon*. (Cantor 1887–88: 401, trans. in Jané 1995: 393)

We are now in a position to lay down one central feature of the concept of set:

> **Unity of Sets.** A set is a *unity*, i.e. a single object, over and above its members.

In particular, a set is a single object bearing a certain distinguished relation – the containment relation (the converse of the membership relation) – to the objects a class as many *aa* comprises. Note, however, that we are making no presumption that to each class as many *aa* there corresponds the set of all the *aa*.

The fact that sets are unities helps distinguish between sets and pluralities. However, it fails to distinguish between sets and what philosophers call *fusions*. A fusion of some entities, in the philosopher's sense, is the same as the *mereological sum* of those entities – that is, the sum of those entities considered as parts of the whole that consists of them.[2] Thus, *a* is a fusion of *b*, *c* and *d* if *b*, *c* and *d* are parts of *a* and every part of *a* shares a part with *b*, *c* or *d*. Hence, a dog is the fusion of the *molecules* which make it up, but it is also a fusion of the *cells* which make it up.

Now some philosophers claim that a fusion is nothing over and above its parts, and is instead *identical* to them. On this view, known as *composition as identity*, a fusion is identical to a plurality – namely the plurality of things that make it up – but is nonetheless a genuine, individual entity.[3] We could try to rephrase Unity of Sets

---

[2] Mereology, the study of the parthood relation, was initiated, in its contemporary form, by Leśniewski (1916). For a historical introduction, see Simons 1991.

[3] The modern debate on whether composition might be identity begins with Baxter 1988. For a contemporary overview including some recent contributions, see Cotnoir and Baxter 2014.

Table 1.1 *Comparing pluralities, fusions and sets*

	Unity	Unique Decomposition
Plurality		✓
Fusion	✓	
Set	✓	✓

so that it requires a set to be *only* an individual entity, so that it cannot be identical to any plurality.[4] But even then, it would remain the case that most philosophers *reject* composition as identity and take instead a fusion to be a single entity different from its parts.

Thus, Unity of Sets cannot serve to distinguish sets from fusions. Instead, what distinguishes sets from fusions is that fusions need not have a *unique decomposition*: as we pointed out, a dog can be decomposed into both molecules and cells. Hence, there is more than one list of things such that they are parts of the dog and their sum *is* the dog. By contrast, sets do have a unique 'decomposition' into members: for any set a, there is one and only one list of things such that they are members of a and their union is a. Thus, whilst fusions need not uniquely decompose into parts, sets do uniquely decompose into members.

This gives rise to another central feature of the concept of set:

> **Unique Decomposition of Sets.**   A set has a *unique decomposition*.

Unlike fusions, therefore, sets have a unique decomposition. But pluralities do too: given a plurality aa of objects, there is only one list of things that are among the aa and such that their union is aa.[5] The situation is summed up in Table 1.1. We have thus located two features belonging to the concept of set: the fact that sets are the result of collecting a plurality into a unity and the fact that this unity has a unique decomposition into members. These features, however, do not suffice to distinguish sets from other ways of collecting pluralities into unities. To deal with such cases, we first need to say something more general about concepts and introduce the notion of a criterion of identity.

---

[4] Potter (2004) takes it to be part of the concept of a fusion that a fusion is nothing over and above its parts, and cites Lewis (1991) to this effect. But note that Lewis is defending a specific view of what fusions are, namely composition as identity.

[5] At least, this is the received view on the matter. Ted Sider (2007) has argued that if a strong form of composition as identity holds, then pluralities do not satisfy Unique Decomposition either.

1.3 Criteria of Application and Criteria of Identity

Thus far we have been talking in a rather unreflective manner about the concept of set. What concepts are is a controversial matter, one on which we need not take a particular stand in this book. It is typically agreed, however, that concepts are associated, one way or another, with what Michael Dummett (1981: 73ff.) calls a *criterion of application.*

Roughly speaking, a criterion of application for a concept C tells us what objects fall under C – to which objects C applies. Consider [$x: x$ is a bachelor]. This concept seems to be associated with the following criterion of application:

$$\text{'bachelor' applies to } x \text{ iff } x \text{ is unmarried and } x \text{ is a man.}[6] \qquad (1.1)$$

Thus, someone who possesses the concept of bachelor will, in normal circumstances, be willing to apply 'bachelor' to an object just in case they take that object to be an unmarried man. (This account of concept possession would need to be refined, but it will do for current purposes.)

Now there is a sense in which the criterion of application for a concept is *constitutive* of that concept: the meaning of 'bachelor' seems to be determined, at least in part, by (1.1). It is perhaps worth stressing that a criterion of application for a concept need not be that in terms of which concepts are explained: someone might accept that a concept has a criterion of application whilst explaining concepts in terms of a variety of other notions such as, e.g., mental representation or inferential role. What matters is that concepts are typically taken to be associated with criteria of application, which are partly constitutive of that concept.

But what about the concept [$x: x$ is a set]? If people do possess the concept of set, then this concept must be associated with a criterion of application. We shall shortly return to the question of what this criterion of application might be. For the time being, we need to notice that besides a criterion of application, certain concepts also have what Dummett calls a *criterion of identity.*

A criterion of identity specifies the conditions under which some thing x falling under a concept C is the same as another thing y, also falling under C.[7] It was Gottlob Frege who first pointed out in *Die Grundlagen* (1884: §54) that only certain concepts are associated with a criterion of identity.

Consider the concept [$x: x$ is red]. Someone who possesses this concept is, in the majority of cases, able to tell whether 'red' applies or fails to apply to a certain

---

[6] Here and throughout, I use 'iff' as an abbreviation of 'if and only if'.

[7] Of course, each thing can only be identical to *itself,* so if that is the case it is not really *another* thing. Similarly, often people ask whether two things are identical, where clearly, if they are, they are not really *two* things. I take it to be clear enough what is meant when using locutions of this kind, and shall therefore indulge in them.

object: she will in general be able to determine when it is correct to say 'That is red'. But there seems to be nothing in the meaning of 'red' which enables one to determine whether two red things are the same or not. Thus, [x : x is red] is associated with a criterion of application but not with a criterion of identity. This highlights a contrast with concepts such as [x : x is a table]: not only is someone who possesses this concept capable of telling, in many situations, whether an object is a table or not, but they are also able to tell whether a table x is the very same table as the table y. [x : x is a table] is associated with a criterion of identity as well as a criterion of application.

The discussion so far suggests that a *criterion of identity* should specify when a thing x (falling under concept C) is identical to a thing y (also falling under C) in terms of a relation Φ holding between x and y. Formally, this gives rise to the following:

$$\forall x \forall y (K(x) \land K(y) \rightarrow (x = y \leftrightarrow \Phi(x,y))), \qquad \text{(CI)}$$

where '$K(x)$' expresses x *falls under* C. Note that since identity is an equivalence relation, the embedded relation Φ must be an equivalence too.

A notorious example is Donald Davidson's (1969) criterion of identity for events, according to which two events are the same just in case they have the same causes and effects:

$$\forall x \forall y (\text{Event}(x) \land \text{Event}(y) \rightarrow (x = y \leftrightarrow (\forall z(z \text{ is a cause of } x \leftrightarrow$$
$$z \text{ is a cause of } y) \land \forall z(z \text{ is an effect of } x \leftrightarrow z \text{ is an effect of } y))). \quad \text{(CIE)}$$

However, some philosophers have argued that not all criteria of identity conform to (CI). Timothy Williamson (1990), for instance, distinguishes criteria of identity of this kind – which he calls *one-level* criteria of identity – from criteria of identity of the following form:

$$\forall \zeta \forall \theta (\S(\zeta) = \S(\theta) \leftrightarrow \Phi(\zeta,\theta)), \qquad \text{(FCI)}$$

where $\S$ is an operator taking items of type ζ and θ to objects. As in the case of (CI), Φ must be an equivalence relation because identity is.

Williamson calls criteria of the form (FCI) *two-level* criteria, since they specify when an object $\S(\zeta)$ is identical to an object $\S(\theta)$ in terms of a relation between entities ζ and θ which are in principle distinct from $\S(\zeta)$ and $\S(\theta)$. Thus, whilst (CI) provides a criterion of identity for the objects falling under a certain concept in terms of a relation Φ holding between these very same objects, (FCI) provides a criterion of identity for objects of a certain kind in terms of a relation holding between entities which may belong to *another* kind.

Two-level criteria are sometimes called *Fregean*, since in *Die Grundlagen* and elsewhere Frege offered many examples of criteria of this kind. The first he considers has now become the paradigmatic example of a two-level criterion:

$$\forall x \forall y (D(x) = D(y) \leftrightarrow x \parallel y). \tag{Dir}$$

(Dir) states that any two lines have the same direction if and only if they are parallel, and is a criterion of identity for *directions*. But the relation $\parallel$ in terms of which (Dir) is formulated is a relation between *lines*, and this makes it clear that it is a two-level criterion. Within the neo-Fregean programme in the philosophy of mathematics (see, e.g., Wright 1983; Hale and Wright 2001), two-level criteria are also known as *abstraction principles*.

Philosophers disagree over the status of the relation between one- and two-level criteria. Williamson (1990; 1991), for instance, contends that neither type is reducible to the other, whilst E. J. Lowe (1989; 1991) claims that two-level criteria can be reduced to one-level criteria and, as a result, the latter are more fundamental. We do not need to enter this debate here, but take the opportunity to note that we will mostly be concerned with one-level criteria, although two-level criteria will also be discussed (see, in particular, Section 1.7 and Chapter 5).

Philosophers also disagree over what *kind of principles* criteria of identity are (see Horsten 2010 for an overview). On the face of it, identity criteria specify what it *is* for two things falling under a certain concept to be identical. As such, they appear to be *metaphysical* principles. However, it also seems to be the case that, just like criteria of application, criteria of identity partly determine the *meaning* of the term they are associated with. In this sense, one might regard them as *semantical* principles. Finally, it seems possible to use criteria of identity to *find out* whether two things falling under a certain concept are in fact one and the same: according to (Dir), I can find out whether the direction of line *a* is the same as *b*'s by finding out whether *a* and *b* are parallel. This aspect of identity criteria – no doubt responsible for their name – seems to make them *epistemic* principles.

The central disagreement, however, does not seem to concern so much whether identity criteria have the aforementioned roles, but rather which role is the *primary* one. Again, we do not need to adjudicate the matter here, since nothing we shall say hinges on the outcome of this dispute: all we shall assume is that identity criteria do have the roles in question.

1.4 Extensionality

Is [x : x is a set] associated with a criterion of identity? It is typically thought that it is, and the reason offered is that we want to be able to count the members of a set: since sets can themselves be members of some sets,[8] it follows that we should be able to count sets themselves. And as Frege (1884: §54) observed, in order to count the objects falling under a concept C we need to be able to tell when any two such objects are in fact one and the same.

Frege's point is well taken: if we want to count the objects falling under a certain concept, it is crucial that we should count each object once, and only once. And that is why if we want to count sets, [x : x is a set] needs to be associated with a criterion of identity (see Dummett 1981: 546–549): such a criterion is required to avoid double-counting.

So when is it that two sets are identical? The standard view on the matter is that the identity conditions for sets are given by the Axiom of Extensionality, which asserts that if two sets have the same members, they are identical:

$$\forall x \forall y (\text{Set}(x) \wedge \text{Set}(y) \to (\forall z(z \in x \leftrightarrow z \in y) \to x = y)). \qquad \text{(Ext)}$$

We will sometimes be concerned with theories which only deal with sets. In that case, the antecedent of (Ext) is unnecessary, and the Axiom of Extensionality takes the following form:

$$\forall x \forall y (\forall z(z \in x \leftrightarrow z \in y) \to x = y). \qquad \text{(Ext}^*)$$

Now since

$$\forall x \forall y (x = y \to \forall z(z \in x \leftrightarrow z \in y))$$

is a theorem of first-order logic with identity, Extensionality delivers the following extensional criterion of identity for sets:

$$\forall x \forall y (\text{Set}(x) \wedge \text{Set}(y) \to x = y \leftrightarrow (\forall z(z \in x \leftrightarrow z \in y))). \qquad \text{(ECS)}$$

We said earlier that criteria of identity are standardly taken to play various roles. One role is that of expressing, more or less directly, aspects of the nature of the objects falling under the concept with which they are associated. Another role is that of partly determining the meaning of the term they are associated with.

The extensional criterion of identity conforms to what we said on this score. For this criterion enables us to distinguish sets from other entities which in some sense collect a plurality into a unity and which do have a unique decomposition,

---

[8] If they weren't – and I hope I am allowed this counterpossible – then the strength of set theory would be so diminished as to compromise most of its use.

Table 1.2 *Comparing pluralities, fusions, sets and properties*

	Unity	Determinate Constituency	Extensionality
Plurality		✓	✓
Fusion	✓		?
Property	✓	✓	
Set	✓	✓	✓

such as *properties* (understood as objective correlates of predicates[9]). Consider, for instance, the property of being a human being and the property of being a feather-less biped. Anything which has the former property also has the latter property, and vice versa; yet we want to say that the two properties are distinct.[10] For a theory to be a theory of *sets*, on the other hand, sets have to be identical if they have the same members.

Thus, Extensionality is part of the concept of set: we are trying to give a theory of objects which fall under a concept such that the objects which fall under it are extensional. As George Boolos (1971: 28) put it, 'a theory that did not affirm that the objects with which it dealt were identical if they had the same members would only by charity be called a theory of *sets* alone' (see also Potter 2004: 33).

Incorporating this fact into Table 1.1 we obtain Table 1.2. The question mark in the row concerning the concept of a fusion signals the fact that the question of whether fusions are identical if they have the same proper parts is still disputed.[11]

The three features that, we have seen, belong to the concept of set and help us distinguish sets from pluralities, fusions and properties are nicely summed up by Oliver and Smiley:

a set is one object related to many objects in a special way. The many determine the one, in the sense that for any given objects there is at most one set of which they are members [...] A special feature of it is that the one determines the relevant many – a set determines which are the objects we call its members. (Oliver and Smiley 2012: 248)

---

[9] And thus not to be confused with what in the philosophical literature are called *natural properties*. These are things which are in part discovered by science and carve reality at the joints, to use an expression from Lewis 1983, which is also the *locus classicus* for the distinction. Properties in the sense at issue here are sometimes also called *concepts*. In Section 6.6 we will see that whilst some theorists take properties in this sense to be objects, others do not. This might provide another way of distinguishing between sets and properties. However, it does not affect the point that sets, unlike properties, are extensional entities.

[10] It is not being suggested that it is *incoherent* to construct properties as extensional entities. The suggestion, rather, is that we have *a* notion of property for which Extensionality does not seem to hold. Gödel, who seems to prefer the term 'concept' over the term 'property', for one, agrees: 'two concepts which apply to the same things are often different. Only concepts having the same meaning [intension] would be identical' (the passage is reported in Wang 1996: 275).

[11] See, for instance, the discussion between Varzi (2008) and Cotnoir (2010).

1.5 What Conceptions Are

Thus, we seem to have the concept of set, and this concept seems to be associated with a criterion of identity and a criterion of application. But what about *conceptions* of set? What are they? And why do we need them if we already possess the concept of set? These questions are of the utmost importance for the problems addressed in this book, and we will focus on them in the remainder of this chapter.

Let us start by considering one example from a domain of inquiry where conceptions are taken to play an important role. The example is intended to bring to light the following phenomenon: two rational, competent English speakers can possess a certain concept and yet disagree as to whether the concept applies to a certain thing – and this disagreement seems to be, in an important sense, *conceptual*.

The example concerns the concept $[x: x$ is fair$]$ and two speakers, Jane and Susan. Jane and Susan, we can assume, possess the concept of fairness: they use it in their lives all the time, and the community of speakers recognize them as competent users of the term 'fair'. Nonetheless, we can easily imagine scenarios in which Jane and Susan disagree over whether a certain action or decision is fair. Consider, for instance, the following case:

> **Mary.** Mary is going to be rewarded by her company for her successful work on a certain case. Jane and Susan, however, disagree over whether this reward is fair: Jane thinks that Mary's work has, in fact, entirely been carried out by Mary's colleague Marianne, whilst Susan is persuaded that it has not.

In this scenario, Jane and Susan disagree over whether 'fair' applies to the company's decision, but the disagreement seems to be due to their having different views over *what the facts of the matter are* – over whether Mary has in fact carried out the work for which she is going to be rewarded.

But now consider the following scenario:

> **Jill.** Like Mary, Jill is going to be rewarded by her company for her work on a certain case. Jane and Susan, however, disagree over whether this decision is fair: according to Jane, Mary should be rewarded because it is fair to reward employees depending on their contribution, whereas according to Susan it is not. For Susan, a company should reward its employees depending on their efforts in their work, regardless of its outcome.

In this case, the disagreement between Jane and Susan is not over what the facts of the matter are. Rather, the disagreement concerns the criterion of application for 'fair' – it seems to be what one might term a *conceptual* disagreement. But now recall that it seems reasonable to assume that both Jane and Susan *possess* the concept of fairness. It seems that their disagreement does not offend against the meaning of 'fairness'. We can say – and political philosophers (e.g. Rawls 1971: 5) sometimes do say – that although both Jane and Susan have the concept of fairness, they have different *conceptions* of it.[12]

This way of putting the matter suggests the following characterization of what a conception is:

> **Conception.** A *conception* of C, where C is a concept, is a (possibly partial) answer to the question 'What is it to be something falling under C?' which someone could agree or disagree with without being reasonably deemed not to possess C.

The idea is that certain features of our use of, say, the term 'fairness' are central to our understanding of the term, whereas certain other features are not. In the case of set theory, as we have seen, it is now thought to be central to our use of the term 'set' that sets are extensional; the same does not apply, for instance, to the statement that every set has a powerset (where, as usual, the powerset $\mathcal{P}(a)$ of a set a is the set of all subsets of a):[13] somebody who disagreed with the latter statement would not be reasonably deemed not to possess the concept of set. The features that are central to our understanding of 'set' are those that determine whether we count somebody as possessing the *concept* of set; those that are not so central, but still constitute some kind of answer to the question 'What is a set?', are those that determine whether we count somebody as possessing a particular *conception* of set.

It might be objected that our characterization of what a conception is rests on a distinction between those uses of a term which are central to our understanding of it and those which aren't. And, the objection goes, this distinction presupposes a distinction between analytic and synthetic truths, a distinction which W. V. Quine (1951a) has shown to be spurious.

It would take us too far afield to enter the vexed debate concerning the analytic/synthetic distinction, but, fortunately, we need not do so. For Quine's attack

---

[12] It is perhaps worth stressing that in saying this, we are not saying that this is the way the terms 'concept' and 'conception' are normally used. For instance, people often use the term 'concept' in roughly the way we are using the term 'conception', for example when they say, in the course of a discussion, that they have a different concept of fairness than ours. Our aim is to draw a distinction which will prove useful in the course of the book.

[13] We say that a set a is a *subset* of a set b (in symbols: $a \subseteq b$) iff every member of a is also a member of b.

on the distinction is largely directed to the distinction being a *sharp* one, whereas our account does not require the distinction between central and not central uses of a term to be sharp. In fact, it seems quite clear that it is not, and there will often be uses of a term such that it is not clear whether someone who contravenes them would be deemed to have not understood the concept in question or to have disagreed about certain features of the objects falling under that concept. But, to repeat, although the distinction between concepts and conceptions is not a sharp one, all we need for our purposes is that there is such a distinction, and that there are clear-cut cases of features of the objects falling under a concept C that are part of C and features that are not.

Nor does the distinction need to be *fixed*: it is compatible with what we have said so far that there can be variation in the range of uses of a term which are central to our understanding of it and those which are not. There may well be uses of a term which were previously associated with a certain *conception* of C but are now associated with the *concept* of C. Indeed, this seems to be the case for the Axiom of Extensionality. Earlier, we said that Extensionality is built into our concept of set in the sense that, in laying it down as an axiom, we make it clear that it is an extensional conception of collection that we are interested in, and that this serves to distinguish sets from entities such as properties. It seems that, now, there is no such thing as an intensional conception of *set*. However, things were different in the second half of the nineteenth century, when, in the early days of set theory, there was a lively debate between intensionalist and extensionalist approaches, to collections and logic more generally.[14]

1.6 What Conceptions Do

So what form can a conception take? The characterization offered in the previous section is compatible with the possibility of a conception having a variety of forms. For instance, it is compatible with the possibility of conveying a conception by providing an account of the nature of the things falling under a concept C – of what being a set consists in – or by specifying necessary and sufficient conditions for being C. We will see examples of this in due course.

But what is a philosopher (or a mathematician, or a logician) doing when she puts forward a conception of set? Here, again, there are various possibilities, which have to do with how one conceives of the relation between conceptions of a concept C and C itself.

---

[14] Hamacher-Hermes 1994 is a detailed account of the controversy. Ruffino (2003) argues that Frege's distinction between sense and reference is an attempt to steer a middle way between intensionalists and extensionalists.

Sharpening

The first possibility is that in offering a conception of set a philosopher might be attempting to *sharpen* the concept of set. That is, in general, by providing a conception of C, one attempts to reduce the range of cases in which two speakers can disagree over whether C applies to a certain thing or not without being deemed not to possess C.

Let us expand on this point and, to this end, introduce some terminology. Call the *range of application* of a concept the class of objects to which a given concept applies. So, for instance, the class of books is the range of application of the concept *book*. And call the *range of disapplication* of a concept the class of things to which the concept *disapplies* – where 'disapplies' is used as an antonym of 'applies'. Now in presenting examples of criteria of application and criteria of identity, I have not explicitly highlighted one important fact: a criterion of application and a criterion of identity for a concept need not settle all questions concerning, respectively, whether the concept applies or disapplies to a certain thing and the identity between objects falling under that concept.

Let us consider two examples to illustrate this point, starting from the second case. Take the concept $[x : x$ is an event]. We saw earlier that Davidson argued that two events are identical just in case they have the same causes and effects. Another criterion of identity was offered by Jaegwon Kim (1973), who took events to be the exemplification by a concrete object (or objects) of a property (or relation) at a certain time. His proposal – with a few qualifications which can be safely ignored here – is as follows:

$\forall x \forall y$ such that x and y are events, $x = y$ iff the constitutive

object(s), constitutive property or relation and constitutive time of

$\qquad\qquad\qquad\qquad\qquad x$ are identical to those of y.   (1.2)

It has been argued (see, e.g., Cleland 1991: 231) that this criterion does not settle all identity questions concerning events because we can conceive of events which do not involve physical objects – for instance, following Peter Strawson (1979: 59–86), we can make sense of a spaceless world of sounds, in which booms, bangs and so on constitute *bona fide* events. Whether Strawson's 'No-Space World' is indeed possible or not, the example shows how a criterion of identity for a concept C might not decide all questions of identity concerning the objects falling under C.

What about the first case? Let us start by considering the concept $[x : x$ is a smidget] (Soames 1999), associated with the following criterion of application:

'smidget' applies to x if x is greater than four feet tall.   (1.3)

'smidget' disapplies to x if x is less than two feet tall.   (1.4)

Clearly, [x : x is a smidget] is such that its range of application and its range of disapplication do not exhaust all possibilities: if x is three feet tall 'smidget' neither applies nor disapplies to it. Now [x : x is a smidget] is an artificially constructed example, and one might doubt that in fact such a concept could be part of our language. Clearly, there are readings of 'could' and 'our language' on which it is true that 'smidget' could not be part of our language: we bump into things that are three feet tall all the time, and perhaps this concept would be rather useless. But it is not very hard to think of practices in which *what matters* is just whether an object is greater than four feet tall or less than two feet tall. In such a case, the concept [x : x is a smidget] might well become entrenched in the practice.

In any case, it is plausible that we do have concepts whose range of application and range of disapplication do not exhaust everything there is. That this is so was argued by Friedrich Waismann. He writes:

Suppose I have to verify a statement such as 'There is a cat next door'; suppose I go over to the next room, open the door, look into it and actually see a cat. Is this enough to prove my statement? [. . .] What [. . .] should I say when the creature later on grew to a gigantic size? Or if it showed some queer behaviour usually not to be found with cats, say, if, under certain conditions it could be revived from death whereas normal cats could not? Shall I, in such a case, say that a new species has come into being? Or that it was a cat with extraordinary properties? (Waismann 1945: 121–122)

Waismann offers the following diagnosis of the situation:

The fact that in many cases there is no such thing as a conclusive verification is connected to the fact that most of our empirical concepts are not delimited in all possible directions. (Waismann 1945: 122)

Most of our empirical concepts, Waismann says, display what he calls *open-texture*. Translated in our terminology, to say that a concept displays open-texture seems to amount to saying that the range of application and the range of disapplication of a concept do not exhaust all possibilities.

Notice, however, that Waismann's focus in the quoted passage is on *empirical* concepts, whereas we are concerned with the concept of set, which, arguably, is not an empirical concept. Does Waismann's point apply to non-empirical concepts as well?

Stewart Shapiro (2006a) has argued that what Waismann says is true for at least one concept that occupies a prominent role in pure mathematics, namely the concept of *computability*. According to him,

in the thirties, and probably for some time afterward, [the pre-theoretic notion of computability] was subject to open-texture. The concept was not delineated with enough precision to decide every possible consideration concerning tools and limitations. (Shapiro 2006a: 441)

But, Shapiro continues, the mathematical and conceptual work carried out by Alan Turing (1936) and the subsequent efforts by the founders of computability theory – Church, Kleene, Post, to name a few – served to sharpen $[x : x$ is computable$]$ into what is now known as the concept of *effective* computability.

Needless to say, Shapiro's suggestion accords with our account of the concept/conception distinction. Using our terminology, the situation can be described as follows. The concept of computability – the *pre-theoretic notion*, as Shapiro put it – displayed open-texture. To be put to mathematical use, the concept needed to be sharpened by putting forward a *conception* of computability, and various candidates presented themselves: computability as effective computability – the conception articulated by Turing and other logicians mentioned above and which abstracts from 'medical' limitations such as finitude of memory or of working space – but also, for instance, computability as *practicable* computability – which does not idealize away from considerations of feasibility and the like (see, e.g., Parikh 1971). Eventually, the mathematical community settled for computability as effective computability, on the basis of, *inter alia*, considerations about the interest and fruitfulness of this notion.

Can the same thing be said about the concept of set? That is to say, does the concept of set, as used in ordinary parlance – with the proviso that we are focusing on *sets* rather than *pluralities* – present open-texture? It is not implausible to think that it does. That is, it is not implausible to think that the concept $[x : x$ is a set$]$ does not settle all questions concerning the extension of 'set'. For it is true that there are some paradigmatic examples of the *use* of the concept of set, which settle *some* questions concerning its extension. For instance, it is quite likely that somebody who refused to say that the extension of 'set' includes the collection of the books that I own would be deemed not to possess the concept of set. One could go as far as saying that the extension of 'set' comprises any finite collection of individuals. But this is a far cry from the concept of set settling all possible cases.

To make things more concrete, consider the putative criterion of application of the concept $[x : x$ is a set$]$ embodied by the following two principles:

$$\text{'set' applies to } x \text{ if } x \text{ is a finite collection of individuals.} \quad (1.5)$$

$$\text{'set' disapplies to } x \text{ if } x \text{ is an individual.} \quad (1.6)$$

This does not tell us whether the concept of set applies or disapplies to, say, the set of all things – where these include *sets themselves* – or sets of the form $x = \{x\}$.

This picture of the situation fits well with the idea that a conception of set is an attempt to sharpen the concept of set. For the idea is then that in providing an answer to the question 'What is a set?' which someone could agree or disagree with without being deemed not to posses the concept of set, a conception of set tells us more about the range of application or the range of disapplication of a concept of set than is embodied in its constitutive principles. In other words, on this view conceptions of set are attempts to reduce the open-texture presented by the set concept.

In addition, providing a conception of set can also enable us to say more on the criteria of identity for sets in cases in which Extensionality does not settle all identity questions about sets. For, as we will see in Chapter 7, whilst in the presence of the so-called Axiom of Foundation (to be introduced in the next chapter), Extensionality suffices to fix the identity conditions of all sets, this need not be the case for theories which do not include that axiom. And the conception of set we will discuss in that chapter indeed leads to more stringent identity criteria for sets than those embodied by Extensionality.

Analysing

So far we have followed Waismann on empirical concepts and Shapiro on the concept of computability and taken the concept of set not to be fixed in all possible directions: there are objects such that the criterion of application for [x : x is a set] does not decide whether 'set' applies to them or disapplies to them. As a result, what conceptions do on this view is sharpen the concept of set.

This picture of what conceptions of set do is meant to be supported by the fact that two competent speakers of English can agree or disagree over whether, say, the set of all sets belongs to the range of application of 'set' without being deemed not to possess the concept of set. The underlying assumption is that this fact about our practices shows something about the *concept of set itself*: it shows that the criterion of application for [x : x is a set] does not settle the question whether the set of all sets belongs to the range of application for 'set'. That is to say, we have concluded that the concept of set is not fixed in all possible directions on the basis of the fact that two competent speakers may disagree over what the criterion of application for [x : x is a set] is.

It has been argued by some, however, that this inference is mistaken: a concept can be fixed in all possible directions even though ordinary speakers can reasonably disagree over whether the term expressed by that concept applies or disapplies to a certain thing whilst agreeing on matters of fact. For instance, this is a crucial aspect of Williamson's (1994) position with regard to vagueness.

Williamson argues that terms such as 'red' are fixed in all possible directions even though native speakers of English might feel that it would be wrong for them to assert or deny that a certain particular shade is red. Yet, Williamson argues, our concept [x: x is a red] is fixed to the extent that it determines whether 'red' applies or disapplies to that shade of red – but sometimes we just cannot know, even in principle, which way. In other words, the union of the range of application and the range of disapplication of 'red' is the totality of all things.

Clearly, on this view, conceptions of set cannot be seen as *sharpening* the concept of set. So what do they do? The natural suggestion is that what they do is spell out in greater detail features of the concept of set that are already, as it were, written into that concept. What is crucial, according to this view, is that what is written into a concept can outstrip our ordinary use of such a concept. In other words, the idea is that what a certain conception of set does is tell us more about features of the concept of set which, though written into it, do not lead us, when denied by someone, to say that that someone does not possess the concept of set.

The view in question seems to have been held by Kurt Gödel (see, e.g., 1944; 1947; 1972), who spoke of *perception* of concepts. According to Gödel, the reason why we can begin with a vague concept and then find a sharp concept to correspond to it faithfully is that the sharp concept was there all along. What happened is that, initially, we did not perceive it clearly. Gödel was wont to liken the situation with ordinary cases of perception: in such cases, it is possible to perceive an object vaguely in the sense that, for instance, we do not see it clearly from a distance; but this does not mean that the object itself is not sharp.

Replacing

Let us say that a concept is *classically inconsistent* (or simply *inconsistent*) iff its constitutive principles are such that they lead to a contradiction in classical logic. Typically, both those that think of conceptions of set as sharpenings and those that take them to be analyses agree that the set concept is consistent. Although this is probably the dominant view, it is by no means the only one available, and some philosophers have claimed that [x: x is a set] is an inconsistent concept (see, for instance, Priest 2006a: 28–30; Woods 2003: 156). According to their view, the range of application of our ordinary concept of set is already sharp and therefore fixed in all possible directions. However, the view continues, the criterion of application associated with this concept is the one described by the naïve conception of set, to be introduced in the next section. As a result, the concept of set is inconsistent.

Different conclusions have been drawn from the claim that the concept of set is an inconsistent one. Some have suggested that we should stick with the inconsistent concept and use it within the context of a logic which makes inconsistencies tolerable. We will be concerned with such views in Chapter 4. Others have argued that we should modify the set concept so as to eradicate the inconsistency whilst at the same time preserving as much of it as possible or maintaining the features of it that make it possible for the concept to have the role that it was meant to play in our practices. Quine's position, to be discussed in Chapter 5, might be taken as falling under this heading. A related or perhaps identical view – depending on how one regards the result of modifying a concept – holds that the inconsistency of the set concept means that we should *replace* it with another, consistent concept. On this view, conceptions of set are proposed *replacements* for the original but inconsistent concept. The fact that [x : x is a set] is a concept whose constitutive principles give rise to inconsistency does not mean that we should stick with it. This is intended to be in analogy with the way we allegedly deal with others ordinary concepts when they are marred with inconsistency: we replace them with consistent, scientifically sound concepts.

A view of this kind has been defended by Kevin Scharp (2013) in the case of truth. According to him, [x : x is true] is an inconsistent concept, which gives rise to a series of paradoxes such as the Liar. This, however, does not mean that we should stick with the original, inconsistent concept of truth. Rather, we should replace it with a different concept that can play a similar role in our practices as the one the original concept of truth was meant to be playing.[15]

Finally, one could have reasons other than inconsistency for wanting to replace a concept: inconsistency is just one feature that might lead one to deem a concept unable to fulfill the tasks we want to use it for. Using again an example from political philosophy, Sally Haslanger (2000; 2012) has argued that the current concepts of *gender* and *race*, though consistent, are problematic in that they perpetuate inequalities. These concepts, says Haslanger, should be replaced using what she calls *ameliorative analyses*, which are, in effect, proposed replacements of the original concept. This can be regarded as a case of so-called *conceptual engineering* (see Cappelen 2018), but note that the approach according to which the concept of set should be sharpened (rather than replaced) could also be considered as an instance of conceptual engineering.

---

[15] According to Scharp, it is in fact two concepts – [x : x is true-in] and [x : x is true-out] – that the ordinary concept of truth should be replaced with.

1.7 What Conceptions Are For

Thus, a conception of set can be taken as doing various things: providing a sharpening of the concept of set; offering an elucidation of aspects of the concept which are supposed to be, as it were, already written into the concept; providing a possible replacement for the inconsistent concept of set. We will not adjudicate between these different views of what conceptions of set do. Rather, we will typically cast the discussion of conceptions of set in terms of the conceptions-of-set-as-sharpening approach, but what we will be saying can easily be rephrased in terms of the other two approaches.[16]

But what is the *purpose* of offering a conception of set in the first place? That is, to say, what are conceptions of set *for*?

'What Is a Set?'

One purpose is, of course, philosophical interest in the question: 'What is a set?'. The language of sets occurs everywhere in mathematics, and indeed set theory is thought by some philosophers to provide a foundation for mathematics (see Section 2.7). Whether one agrees or not, the pervasiveness of set-theoretic language is hard to dispute. If that is true, the question what sets are becomes an important one for a philosopher of mathematics. But, as Oliver and Smiley say, the concept of set (taken to have the features we attributed to it above) 'still only specifies the set to a very limited degree: beyond that it leaves it quite unsettled what sort of thing a set is' (2012: 248). At the same time, it is hard to come up with features of set that can be taken to belong to the concept of set without begging some question about what sets are. However, one may instead talk about features that sets have on a certain *conception* of set.

Now the question 'What is a set?' can be understood in a variety of ways, depending on what one takes oneself to be doing when offering a conception of set. If one takes a conception of set to be a sharpening of the concept of set, then the question is naturally understood as demanding us to provide an account of what being a set *consists in* – where what a set consists in is not, as it were, already written into the concept. If one takes a conception of set to provide an analysis of

---

[16] With the exception, perhaps, of our discussion of attempts to rehabilitate the naïve conception of set in Chapter 4 by changing the logic of set theory. For someone who holds that an inconsistent concept should be replaced by a consistent one would regard any attempt to salvage the naïve conception as hopeless. However, the discussion in this case can be recast as concerning proposed replacements for the package consisting of the inconsistent concept of set and classical logic. One such replacement consists of a new package consisting of the original inconsistent concept and a different logic.

the concept of set, then the question 'What is a set?' can be understood as asking for an account of what is part of the *concept* of set – where there are some aspects of the concept of set that are not transparent to us. Finally, if one regards a conception of set as providing a possible replacement for the concept of set, then the question 'What is a set?' can be taken to be asking for an account of what the replacement to be adopted is like. In any case, providing a conception of set is going to provide an answer to the philosophical question 'What is a set?'.

One might insist that conceptions of set like the ones we will discuss in the remainder of this book still leave it rather unsettled *what sort of thing a set is*, to put it as Oliver and Smiley do. Here an analogy with the concept of composition will be helpful. Peter van Inwagen (1990) asked three related questions concerning *composition*, only two of which need detain us. The first, known as the *General Composition Question*, asks *what it is* for some things xx to compose y, i.e. to form a fusion. The second, known as the *Special Composition Question*, asks *under what conditions* some things yy compose y. To be satisfactory, an answer to the General Composition question must be *reductive*, that is, it must not involve any mereological concepts, or so van Inwagen argues. Van Inwagen goes on to express pessimism about the possibility of providing such an answer, and suggests that attention should instead be directed to the Special Composition Question, to which he offers his own answer.

Turning to our concerns, conceptions of set need not provide a reductive answer to what one may call the *General Set Formation Question*, that is the question of *what it is* for some things to form a set. For it is true that there are conceptions of set which do seem to provide such an answer. One example is the naïve conception of set, which we will encounter shortly and according to which some things form a set if and only if some predicate applies to them (and only them). But there are also examples of conceptions which do not offer a reductive answer to the General Set Formation Question. A case in point is the iterative conception (see Chapter 2), which rests on notions such as the *set of* operation or the cumulative hierarchy, which might be deemed to be set-theoretical.

Nonetheless, even if they do not provide an answer to the General Set Formation Question, conceptions of set do provide *some* answer to the question 'What is a set?'. For instance, as we shall see, the iterative conception takes a set to be an object obtainable by iterating the *set of* operation, which is an answer to the question 'What is a set?'. What is more, as we explained in the previous section, the answer they provide enables one to make substantial progress on what one may call the *Special Set Formation Question*, that is, the question under what conditions some things form a set, i.e. they allow us to say more about the application criterion for 'set'.

Justification

Another purpose a conception of set might be invoked to achieve – and one which has been historically important – is that of *justifying* theories of set. These tend to have two sorts of axioms: axioms of *nature* and axioms of *extent* (see Koellner 2003: 16–17). Axioms of nature are justified on the basis of what the conception takes sets to be and on the basis of what our concept of set is; axioms of extent are justified on the basis of how many sets there are according to the conception. Sometimes, axioms do not fit neatly into either of these categories; sometimes they fit into both. We will see, however, that most of the time the distinction is a useful one.

Now how can a conception of set be used to justify a certain set of axioms? To answer this question, it will be helpful to introduce some terminology. We will say that a conception of set C *sanctions* the axioms of a theory T if and only if the axioms of T are true in the universe of sets as characterized by C, and similarly for the concept of set. Sometimes, we will also say that the axioms of T are true or justified *on* the conception C (as opposed to being true *tout court* or being justified *by* C). Occasionally, we will also say that the axioms of T *follow* from C.

It seems especially apt to say that the axioms of T follow from C when C (or some part thereof) has been formalized and it is shown that the axioms of T can be deduced from the axioms capturing (at least part of) C. For suppose we want to show that the axioms of a theory T are justified on a certain conception C. The standard strategy is to appeal to an intuitive reading of the axioms of T, to what they express, and say that what they express is the case on that conception. We will shortly see an example of this sort of strategy, in connection with the so-called *Comprehension Schema*. A different type of strategy consists in *formalizing* C (or part of it) and showing that the axioms of T follow from the said formalization. We will encounter instances of such a strategy in Chapters 2 and 7.

Clearly, the fact that a conception sanctions certain axioms does not show that these axioms are *true*, since things might not be as the conception says they are. We can talk of a certain axiom being justified on a certain conception without being committed to the axiom being true. For truth, we also require that the conception is *correct*: if a certain conception *is* correct, then, any axiom sanctioned by the conception will be true: thus, we can offer reasons for thinking that the axioms of a certain set theory are true by offering reasons for thinking that they are true on a conception of set and that the conception of set is correct.

So, strictly speaking, it is only when reasons have been given for thinking that the conception C motivating T is correct that we can say that C *justifies* T.[17] Hence,

---

[17] As we will see in Section 2.7, whether C can sanction a theory which allows us to carry out enough set-theoretic constructions might itself be part of the reasons for regarding C as correct.

to argue that a conception of set justifies the axioms of a certain theory, one needs to argue for two claims: that the conception sanctions the axioms of that theory, and that the theory is correct.

Let us give an example. The *naïve conception of set* holds that sets are extensions of predicates, where the extension of a predicate is the collection of all the things to which the predicate applies. The naïve conception is usually taken as sanctioning the axiom of extent known as the *Axiom Schema of (Naïve) Comprehension*, which states that every condition expressible in the language of our theory determines a set. In symbols:

$$\exists y \forall x (x \in y \leftrightarrow \varphi(x)), \tag{Comp}$$

where $\varphi(x)$ is any formula in $\mathcal{L}_\in$ in which x is free and which contains no free occurrences of y. This means that instances of this and other schemata may contain free variables not displayed, also known as *parameters*. Such instances are tacitly understood as being prefaced by universal quantifiers binding these free variables.

Together with the Axiom of Extensionality, which, as we have seen, is typically taken to be true on the concept of set, this gives rise to classical *naïve set theory*. This theory is as simple as it can be: its logic is classical logic; its language – the language of set theory $\mathcal{L}_\in$ – is obtained by adding the two-place relation symbol $\in$ to the language of first-order logic; and we only have two axioms, an axiom of nature and an axiom of extent. Thus, to sum up, the axioms of naïve set theory are justified on the naïve conception, which means that if the naïve conception is correct – which, we will argue, it is not – the axioms of naïve set theory are justified *tout court*.

The naïve conception, then, holds that every predicate determines a set. We can also think of *properties* as having an extension, namely the collection of all the things having that property. If we do so, then we can regard the naïve conception as obtained from two components: the thesis that every condition determines a property, and the thesis that every property has an extension.

These two components of the naïve conception so conceived can be naturally formulated using the language of second-order logic. Proof-theoretically, second-order logic is obtained by adding to first-order logic rules for the second-order quantifiers and an Axiom of (Second-Order) Comprehension, which states that to every open formula with parameters there corresponds a (possibly relational) property or concept:[18]

$$\exists X^n \, \forall x_1, \ldots, x_n \, (X^n x_1, \ldots, x_n \leftrightarrow \varphi(x_1, \ldots, x_n)), \tag{Comp2}$$

---

[18] Often, a second-order version of the Axiom of Choice (see Section 2.1) is also taken to be part of the deductive system, but we will not include it here in view of the discussion in Chapter 5, where the distinction between second-order choice principles will matter.

where X^n is a second-order variable of nth-degree. To simplify notation, I will omit the superscript from now on and let the context disambiguate. The model theory for second-order logic is traditionally carried out using set theory (Shapiro 1991: 70–76). In the *standard* or *full* model theory for second-order logic, the monadic second-order variables are interpreted as ranging over all subsets of the first-order domain, dyadic variables as ranging over all sets of ordered pairs of objects from the domain, and so on. In the *Henkin* model theory, the monadic second-order variables are interpreted as ranging over some (but not necessarily all) subsets of the first-order domain, dyadic variables as ranging over some (but not necessarily all) sets of ordered pairs of objects from the domain, and so on. In this book, we shall restrict attention to the standard model theory for second-order logic.

The thesis that every predicate determines a (one-place) property is now just a special case of (Comp2) and therefore follows from the standard deductive system for second-order logic. It can be stated as follows:

$$\exists X \forall x \ (Xx \ \leftrightarrow \ \varphi(x)). \qquad \text{(M-Comp2)}$$

The thesis that every property or concept has an extension becomes

$$\forall X \exists y \mathrm{Det}(X, y), \qquad \text{(Collapse)}$$

where $\mathrm{Det}(X, y)$ says that X determines y.

The naïve conception of set can, with some historical caution, be traced back to Frege (1893/1903). Frege works using what nowadays would be considered second-order logic and takes second-order variables to range over concepts, which, in his view, are what predicates refer to. According to Frege, to every concept F we can associate a certain object $\mathrm{Ext}(F)$, the *extension* of F, much in the way in which, according to the naïve conception, to every predicate we can associate the set of all and only the things to which that predicate applies. Formally, Frege takes extensions to be governed by the infamous *Basic Law V*, stating that the extension of a concept F is the same as the extension of a concept G if and only if F and G are coextensive. In symbols:

$$\forall F \forall G \mathrm{Ext}(F) = \mathrm{Ext}(G) \leftrightarrow \forall x(Fx \leftrightarrow Gx). \qquad \text{(V)}$$

Note that this is a two-level identity criterion (see 1.3), which specifies the identity conditions of extensions in terms of a relation between concepts.

Consistency

Naïve set theory is inconsistent. It gives rise to a variety of paradoxes, the simplest of which – Russell's Paradox – proceeds as follows:

1. $\exists y \forall x (x \in y \leftrightarrow x \notin x)$ By (Comp)
2. $\forall x (x \in a \leftrightarrow x \notin x)$ Supp
3. $a \in a \leftrightarrow a \notin a$
4. $\bot$
5. $\bot$ 1, 2–4, by $\exists$-elimination.

The paradox shows that the supposition that there is a set of all sets that are not members of themselves – the Russell set – leads to a contradiction, and it does so with the help of prima facie innocuous logical principles.[19]

Moreover, as Øystein Linnebo (2010) points out (although his discussion is framed in terms of pluralities rather than concepts), (M-Comp2) and (Collapse) can also be used to derive Russell's Paradox. For, clearly, by (M-Comp2), there is a Russellian property R which holds of an object if and only if that object does not belong to itself:

$$\forall x (Rx \leftrightarrow x \notin x). \tag{1.7}$$

But by (Collapse), R determines an object r characterized as follows:

$$\exists r \forall x (x \in r \leftrightarrow Rx). \tag{1.8}$$

And (1.7) and (1.8) jointly entail the existence of the Russell set, from which a contradiction can be derived in the usual way.

Finally, in a letter to Frege, Russell (1902) originally presented his paradox as targeting Basic Law V, and indeed one can show that, in the presence of (Comp2) (and indeed (M-Comp2)), (V) is inconsistent:

Proof. By M-Comp2, there exists the Russellian concept F characterized as follows: $[x : \exists G(x = \text{Ext}(G) \wedge \neg Gx)]$. Now suppose that $\text{Ext}(F)$ does not fall under F. Then $\forall G(\text{Ext}(F) = \text{Ext}(G) \rightarrow G\text{Ext}(F))$ and so $\text{Ext}(F)$ does fall under F. But by definition of F, this means that there is a G such that $\text{Ext}(F) = \text{Ext}(G)$ and $\text{Ext}(F)$ does not fall under G. By the left-to-right direction of (V), it follows that F and G are coextensive, which implies that $\text{Ext}(F)$ does not fall under F either. Contradiction. □

Whenever a paradox arises, two questions present themselves: why does the paradox arise? And what should be done? The first question asks for a *diagnosis* of the paradox, the second for a *remedy* – unless, of course, the answer is simply: nothing.

---

[19] Note, in particular, that $\varphi \leftrightarrow \neg\varphi$ is a contradiction even in intuitionistic logic.

Diagnosis

Let us start from the diagnostic component. Our diagnosis is that Russell's Paradox arises because on the naïve conception we have both *indefinite extensibility* and *universality* of the concept of set not belonging to itself. The term 'indefinite extensibility' is due to Dummett, who gives the following characterization of an indefinitely extensible concept:

An indefinitely extensible concept is one such that, if we can form a definite conception of a totality all of whose members fall under that concept, we can, by reference to that totality, characterize a larger totality all of whose members fall under it. (Dummett 1993: 454)

The first to draw attention to the phenomenon of indefinite extensibility is Russell, to whom our diagnosis of the paradoxes is indebted:

[T]he contradictions result from the fact that [...] there are what we may call *self-reproductive* processes and classes. That is, there are some properties such that, given any class of terms all having such a property, we can always define a new term also having the property in question. Hence, we can never collect *all* the terms having the said property into a whole; because, whenever we hope we have them all, the collection which we have immediately proceeds to generate a new term also having the said property. (Russell 1906: 36)

Following Russell and taking a totality of which we can form a definite conception simply to be a set (or class, in Russell's terminology), we can characterize indefinite extensibility and universality of a concept thus:

> **Indefinite extensibility.** A concept C is *indefinitely extensible* iff whenever we succeed in defining a set u of objects falling under C, there is an operation which, given u, produces an object falling under C but not belonging to u.

> **Universality.** A concept C is *universal* iff there exists a set of all the things falling under C.

On our diagnosis, Russell's paradox is an instance of the more general fact that if a concept C is taken to be both indefinitely extensible and universal, contradiction ensues because all the things falling under C form a set u by Universality but we can find an object falling under C and not belonging to u by Indefinite Extensibility – we can *diagonalize out* of u.

Russell's own treatment can be used to give a more formal account of the situation. Let φ be a predicate – a propositional function, in Russell's terminology – and f a function. Then, to say that

$$\forall u (\forall x (x \in u \to \varphi(x)) \to (f(u) \notin u \land \varphi(f(u)))) \qquad \text{(IE)}$$

holds for φ and f is to say that the concept expressed by φ is indefinitely extensible. And to say that the instance of Comprehension

$$\exists y \forall x (x \in y \leftrightarrow \varphi(x)), \tag{U}$$

holds for φ is to say that the concept expressed by φ is universal. But given this, it is easy to see that any concept which is both indefinitely extensible and universal will be an inconsistent concept:

Theorem 1.7.1. (IE) *and* (U) *jointly yield a contradiction.*

Proof. From (U) and (IE) we get, by existential instantiation and universal instantiation, $\forall x (x \in w \leftrightarrow \varphi(x))$ and $\forall x (x \in w \rightarrow \varphi(x)) \rightarrow (f(w) \notin w \wedge \varphi(f(w)))$. It follows that $f(w) \notin w \wedge \varphi(f(w))$. But using the second conjunct and $\forall x (x \in w \leftrightarrow \varphi(x))$ we also have that $f(w) \in w$. $\square$

Note, again, that the inference rules used in the proof appear to be rather innocent ones.

How does Russell's Paradox fit this diagnosis? The idea is that if we take C to be $[x : x \notin x]$, we get a contradiction because on the naïve conception there is a set r of all things that are not members of themselves, and we can find an object falling under the concept $[x : x \notin x]$ but not belonging to r by simply taking this object to be r itself. More formally, we get Russell's Paradox because on the naïve conception (IE) and (U) hold when we let $\varphi(x)$ be $x \notin x$ and $f(x)$ be simply x.

The other set-theoretic paradoxes make use of different choices of f and – predictably, since they concern different concepts – different values of φ.

Making the simplifying assumption that our quantifiers only range over sets, the paradox of the set of all sets is obtained by letting $\varphi(x)$ be $x = x$ and $f(x)$ be $\{y \in x : y \notin y\}$. Russell-style reasoning tells us that $f(x) \notin x$. Thus, we have a recipe which, for any set a, gives a set which cannot belong to a, which is just another way of putting the fact that the set concept is indefinitely extensible. Trouble arises when a already has all sets as members. Since there is a set of this kind on the naïve conception (the universal set $\{x : x = x\}$), contradiction ensues.

Next, we have Cantor's Paradox and the Burali-Forti Paradox, which concern, respectively, the concept of cardinal number and the concept of ordinal number (see Appendix 1.A for a brief introduction to the so-called Frege-Russell definitions of cardinals and ordinals as well as the now standard definitions going back to von Neumann). Cantor's Paradox results from considering a recipe which, for any cardinal κ, produces a cardinal larger than κ. Since the Comprehension Schema implies the existence of a set of all cardinals, we obtain a contradiction. For it is easy to show that if there were such a set, it would have a largest element. More specifically, the required instances of (IE) and (U) are obtained by simply letting f be the powerset operation and $\varphi(x)$ be 'x is a cardinal'. The hard part is to show that (IE) does hold in this case, and here we make use of Cantor's Theorem, which uses certain set-theoretic resources and states that the powerset of a set is larger than the set itself (see Appendix 1.B for details).

The Burali-Forti Paradox goes as follows. Consider the series of all ordinals less than α. This is well-ordered and so its order type is an ordinal number. One can show that this ordinal number must be α itself. Now consider the set of all ordinals. This set too is well-ordered, so its order type is an ordinal number, Ω. However, the series of all ordinals less than Ω also has ordinal number Ω. So we have that the set of all ordinals is isomorphic to a proper initial subset of itself, namely the series of ordinals less than Ω. This is a contradiction. According to the diagnosis offered here, the Burali-Forti Paradox is obtained by taking $\varphi(x)$ to be 'x is an ordinal' and $f(x)$ to be the ordinal number of the well-ordering relation on x.

Finally, to get Mirimanoff's (1917) Paradox concerning the concept $[x : x$ is a well-founded set], put 'x is a well-founded set' for $\varphi(x)$ and $x \cup \{x\}$ for $f(x)$.[20] When a is a set of well-founded sets, $f(a)$ must be well-founded too. So, again, we have a recipe which, given any set of well-founded sets, gives us a well-founded set which is not in that set, and contradiction ensues because the Comprehension Schema entails the existence of a set of all well-founded sets.

Remedy

What about the *remedy*? Depending on one's views about inconsistency, this might consist in a *cure* eradicating the inconsistency or in a *palliative* helping us live with it, and we shall look at views of both kind during the course of the book. But in the case in which what we want is a cure, conceptions of set can help us be confident in the success of the cure: a conception of set can provide us with a reason to think that the axioms of a certain set theory are consistent. Together with a certain diagnosis of the paradoxes, a conception of set may reassure us that the paradoxes will not arise in the theory it sanctions, and this is yet another reason to be interested in conceptions of set.

There are at least two ways in which a conception of set can give us confidence in the consistency of the theory it lies behind. First, the conception can reject one of indefinite extensibility and universality. Since according to our diagnosis, the

---

[20] A relation E on a set a is *well-founded* iff every nonempty $b \subseteq a$ has an E-minimal element, i.e. a $c \in a$ such that there is no $b \in a$ with bEc. A set a is then said to be *well-founded* iff the membership relation on a is well-founded.

paradoxes arise from the interaction of these two features the naïve conception ascribes to the set concept, this provides reassurance that the familiar set-theoretic paradoxes will not arise in the theory. In the course of this book, we will see examples of conceptions that reject indefinite extensibility and ones that reject universality. Second, the conception can provide us with an intuitive model of the theory, which may provide one with reasons for thinking that no contradiction is going to arise in the theory. This, as we shall see, is the case with the iterative conception of set, which appears to provide an intuitive model of the theory it sanctions in the cumulative hierarchy.

The importance of this aim that conceptions of set are sometimes invoked to achieve can be further appreciated by considering the debate between platonists and fictionalists. A platonist can believe that a certain mathematical theory is consistent on the basis of the fact that she has reason to believe that its axioms are *true*. To this extent, the platonist may perhaps ignore the question of consistency and focus on the question whether the axioms of a certain set theory are justified, and, as we saw in the previous subsection, conceptions of set can be used to this end. The fictionalist, however, will want to resist this strategy, since she takes most mathematical axioms to be in fact *false*. For her, no conception of set can be correct – things cannot be as *any* conception of set says they are, since there are no non-trivial mathematical truths at all.

Although they deny that mathematical theories are true, fictionalists typically want to say that consistency is an essential mark of a good mathematical theory. But what reason can fictionalists have for thinking that a certain mathematical theory is consistent? Often, they will say that they believe a mathematical theory T to be consistent because the existence of a model of T follows from S, a set theory which they believe to be consistent: if S implies that T has a model and S is consistent, then T will be consistent too.

But what reason do fictionalists have for thinking that the *background set theory* S is consistent? Various proposals have been advanced, but our discussion suggests that fictionalists can use conceptions to provide the required reason: although no conception of set is correct, if the axioms of a theory S are justified on a conception which appears to involve no contradictions and on which it is not the case that both universality and indefinite extensibility hold, then there will be reason to regard S as consistent.[21]

---

[21] This line of reasoning is developed by Leng (2007: 104–107). According to her, '[w]hile the appearance of consistency does not always imply genuine consistency, our ability to form an apparently coherent conception of a structure satisfying the axioms of a theory should, nevertheless, help to increase our confidence in the consistency of those axioms' (Leng 2007: 106).

1.8 Logical and Combinatorial Conceptions

We conclude the chapter by introducing another important distinction, namely that between *logical* and *combinatorial* conceptions of a collection.

According to the logical conception, a collection is associated to some predicate, concept, or property, which determine its members. As Penelope Maddy puts it:

The logical notion, beginning with Frege's extension of a concept, takes a number of different forms depending on exactly what sort of entity provides the principle of selection, but all these have in common the idea of dividing absolutely everything into two groups according to some sort of rule. (Maddy 1990: 121)

Membership in a logical collection is determined by the satisfaction of the relevant condition, falling under the relevant concept or having the relevant property. Membership is, in a sense, *derivative*: we can say that an object a is a member of a collection b just in case b is the extension of some predicate, concept, or property that applies to a.

The logical conception of a collection is usually contrasted with the *combinatorial* conception. According to this conception, a collection is characterized not by reference to some condition, property or concept, but by reference to its members.

As Shaughan Lavine (1994: 77) noted, this conception arguably goes back to Cantor, who in a letter to Jourdain took a set to be 'defined by the enumeration of its terms'.[22] Similarly, Paul Bernays, to whom the combinatorial conception owes its name, writes that on a combinatorial (or, as he puts it, *quasi*-combinatorial) conception,

one views a set of integers as the result of infinitely many independent acts deciding for each number whether it should be included or excluded. We add to this the idea of the totality of these sets. Sequences of real numbers and sets of real numbers are envisaged in an analogous manner. (Bernays 1935: 260)

In the case of a combinatorial collection, there need not be a predicate that applies to all and only the members of the collection. This is compatible with the fact that in those cases in which such a predicate exists, the collection can be referred to by means of it. Note that on an abundant conception of property, it is trivial to come up with a property had by all and only the members $x_1, x_2, \ldots$ of a combinatorial collection – just take the property $y = x_1 \vee x_2 \vee \ldots$. However, one might require that the relevant property should satisfy certain additional constraints for it to be

---

[22] The letter is reprinted in Grattan-Guiness 1971: 118–119 and translated in Lavine 1994: 98–99.

used to characterize a collection – for instance, we might require that the relevant property should be specifiable by means of some predicate.[23]

Since a combinatorial collection is characterized by reference to its members, membership in the collection is not derivative upon satisfying a certain condition or having a certain property, but is primitive.

The distinction between a logical and a combinatorial conception was arguably an important factor in the dispute over the status of the Axiom of Choice, and we shall return to this issue in the following chapter. For now, note that the naïve conception of set, as standardly formulated, is a logical conception of set. For according to the conception every condition determines a set, and since the conception takes this to be our sole guide to what sets there are, the conception seems committed to the idea that membership in a set is determined by whether the predicate, property or concept associated with the set applies to an object or not. Whether the naïve conception should be regarded as the best way to articulate a logical conception of collection is a question we shall have occasion to return to.

We have seen that the logical conception is potentially more restrictive than the combinatorial conception, since there might be combinatorially obtained collections which cannot be seen as the extension of any specifiable condition (see also Lavine 1994: 77). From a different point of view, however, it is the combinatorial conception that is potentially more restrictive. For, on the face of it, the combinatorial conception does not seem to be committed or even lend some initial plausibility to the thesis that every property determines a collection. The logical conception, by contrast, is naturally associated with this thesis. At least, this thesis is part and parcel of the naïve conception of set, the original and paradigmatic example of a logical conception of collection.[24] However, as we pointed out earlier, the thesis that every property determines an extension leads to paradox in the presence of the thesis that every condition determines a property. It is now time to turn to a conception of set which appears to be consistent.

Appendix 1.A Cardinals and Ordinals

In this appendix, we give the basic definitions of cardinals and ordinals. We focus, in particular, on the differences between the Frege-Russell definition and the

---

[23] Of course, in order for the distinction not to be trivialized, one also needs to have some constraints on what counts as an admissible predicate. For instance, an interesting question is whether the predicate should be specifiable by finitary means or not. For discussion, see Fine 2005: 567–571.

[24] We will discuss logical conceptions which do restrict the thesis in Chapters 5 and 6.

standard definition. A third definition, the so-called *Scott-Tarski* definition, will be given in Section 3.2.

Let us start from the notion of a cardinal number. Intuitively, cardinal numbers are numbers which represent the *size* of collections. The idea is to associate with each set a an object – called the cardinality of a and denoted $|a|$ – which represents the size of a. Now say that sets a and b are *equinumerous* if there is a one-to-one correspondence between them. It was Cantor's insight that a theory of size should be based on the concept of equinumerosity in the sense that we want the following to come out true:

$$|a| = |b| \text{ iff } a \text{ and } b \text{ are equinumerous.}$$

The Frege-Russell definition immediately delivers this. According to this definition, a cardinal is an equivalence class of equinumerous sets. That is to say, the cardinal of a set a is the set of all sets equinumerous with a. This definition is known as the Frege-Russell definition because Frege (1884) took the cardinal number belonging to a concept F to be the extension of the concept *equinumerous with F*, and Russell (1903) took the cardinal number associated with a set a to be the set of all sets equinumerous with a. This definition of a cardinal goes back to Cantor, although how faithful this characterization is to his ideas is disputed (see Hallett 1984: ch. 3).

A similar definition is available for the ordinals. The idea here is that ordinals are numbers which represent the *order structure* of well-ordered collections.[25] So let us write $\langle a, R \rangle$ to denote a set a ordered by the relation R. Analogously to the case of cardinal numbers, we can then define the order-type of a set to be an equivalence class of sets that are isomorphic with respect to R. That is, the order-type of $\langle a, R \rangle$ is the set of all $\langle b, R \rangle$ that are isomorphic to $\langle a, R \rangle$. Ordinals are then just order-types of well-ordered sets, and they will generally be denoted by lowercase Greek letters. In particular, we reserve α, β and λ for ordinals.

The Frege-Russell definition of a cardinal, and the corresponding definition of an ordinal, are rather natural. However, they cannot be used to give a satisfactory development of cardinals and ordinals in the standard system for set theory ZFC, which we will describe in the next chapter. To anticipate, the problem in the case of cardinals is that for any set a other than the empty set, the set of all sets equinumerous with a does not exist in ZFC, because it is unbounded in the hierarchy. Similar considerations apply, *mutatis mutandis* to the case of ordinals.

---

[25] A relation R on a set a is a *total order* of a if it is transitive, antisymmetric and connex (i.e., for every $a, b \in a$ either aRb or bRa). We then say that R is a *well-order* if in addition to being total it also has the property that every non-empty subset of a has a least element with respect to R.

For this reason, a different definition of cardinals and ordinals is used in standard set theory. Let us start with ordinals this time. According to the *von Neumann definition*, an ordinal is just a transitive set well-ordered by membership, where a set a is *transitive* just in case every member of a member of a is also a member of a. We then let $\alpha < \beta$ iff $\alpha \in \beta$. One can then check in ZFC that the ordinals are well-ordered by $<$, that for each α, $\alpha = \{\beta : \beta < \alpha\}$, and that for every α, $\alpha \cup \{\alpha\}$ exists. In the light of the last fact, we define $\alpha + 1$ as $\alpha \cup \{\alpha\}$ and call it the *successor of* α. Furthermore, we let $0 = \emptyset$, $1 = 0 + 1$, $2 = 1 + 1$ and so on. If $\alpha = \beta + 1$, then α is a *successor ordinal*. If α is neither 0 nor a successor ordinal, it is a *limit ordinal* and we have that $\alpha = \sup\{\beta : \beta < \alpha\}$. If it exists, ω is the least limit ordinal. Finally, one can prove the Mirimanoff-von Neumann result that every well-ordered set is isomorphic to a unique von Neumann ordinal. This is a special case of Mostowski's Collapsing Lemma (Jech 2003: Theorem 6.15). The Lemma is provable in ZFC and, as we will see in Chapter 7, in the non-well-founded set theory ZFA. Thus, von Neumann ordinals seem to adequately represent the order structure of well-ordered collections.

Given the von Neumann definition of an ordinal, we can then define cardinals using the *von Neumann cardinal assignment*. It is a theorem of ZFC that every set can be well-ordered. We can then define $|a|$ to be the smallest ordinal equinumerous with a. That is, cardinals are initial ordinals. The finite ordinals $0, 1, 2, \ldots$ and ω are all cardinals. Cardinals are normally denoted by lowercase Greek letters and we reserve κ for cardinals.

The cardinalities of infinite sets that can be well-ordered are also known as *alephs* (in symbols $\aleph_\alpha$). The series of alephs can be defined using transfinite recursion. This is a method of definition in which we define a function on the ordinals by defining what the function does for the case of 0, for the case of successor ordinals, and for the case of limit ordinals (see, e.g., Jech 2003: 22). Let α^+ be the least well-ordered cardinal number greater than α. Then the $\aleph_\alpha$s are defined thus:

$$\aleph_0 = \omega;$$
$$\aleph_{\alpha+1} = \aleph_\alpha^+;$$
$$\aleph_\lambda = \bigcup_{\alpha < \lambda} \aleph_\alpha \text{ if } \lambda \text{ is a limit ordinal.}$$

Appendix 1.B Cantor's Theorem

The standard proof of Cantor's Theorem makes use of certain logical and mathematical resources, which are not available in all the set theories we shall encounter in this book. For this reason, we give the proof here, making explicit a number of mathematical and logical assumptions used in the proof.

Theorem 1.B.1 (Cantor). $|a| < |\mathcal{P}(a)|$.

Proof. It is easy to show that $|a| \leq |\mathcal{P}(a)|$. For, by the definition of partial ordering, this requires the existence of a one-to-one function from a to $\mathcal{P}(a)$, and such a function is given by $x \mapsto \{x\}$. So to show that $|a| < |\mathcal{P}(a)|$, it is enough to show that $|a| \neq |\mathcal{P}(a)|$.

To this end, consider the set $b = \{x \in a : x \notin f(x)\}$ and suppose for *reductio* that there is a function f from a onto $\mathcal{P}(a)$. Then $b = f(c)$ for some $c \in a$. But then $c \in b$ if and only if $c \notin f(c)$ if and only if $c \notin b$. By the transitivity of the biconditional, we have that $c \in b$ if and only if $c \notin b$. Contradiction. Hence $|a| \neq |\mathcal{P}(a)|$. Therefore $|a| < |\mathcal{P}(a)|$. $\qquad\square$

Note that the existence of the set $b = \{x \in a : x \notin f(x)\}$ follows from the Separation Schema, to be introduced in the next chapter (see Section 2.1 and Table 2.1).

2

The Iterative Conception

As long as set theory is thought of
as being about the iterative model,
there will be a temptation for philosophers
to think of the points in this model
as substantial entities related by
some specially distinguished relation,
and to inquire into the metaphysical nature
of these entities and relation.

Allen Hazen (1993: 176)

In the previous chapter, we discussed the distinction between concepts and conceptions, and introduced the naïve conception of set. The main task of this chapter is to present what is known as the *iterative conception of set*.

This conception maintains that sets can be arranged in a cumulative hierarchy divided into levels. Accordingly, we will start from the cumulative hierarchy, and will then move on to illustrate set theories that have been thought to be based on it. Some of these theories are well known, and readers familiar with them will be able to skim through the relevant sections.

According to its proponents, the iterative conception is correct. And according to several of them, this is because there exists a substantial relation of priority or dependence between a set and its members. The second part of the chapter criticizes accounts of the iterative conception that postulate such a relation, and offers an alternative understanding of the iterative conception that dispenses with the said dependence relation. The chapter concludes by outlining a strategy for defending the conception so understood and discussing criteria for evaluating conceptions.

2.1 The Cumulative Hierarchy

In Chapter 1, we offered the following diagnosis of the set-theoretic paradoxes: they arise because on the naïve conception we have both indefinite extensibility and universality of some set-theoretic concept, such as the concept of set and the concept of non-self-membered set. If this diagnosis is correct, it is natural to try to deal with the paradoxes by articulating a conception of set that only retains one of these two aspects. The *iterative conception of set* retains indefinite extensibility and rejects universality.

On the iterative conception, sets are formed in stages. In the beginning we have some previously given objects, the individuals. At any finite stage, we form all possible collections of individuals and sets formed at earlier stages, and collect up the sets formed so far. After the finite stages, there is a stage, stage ω. The sets formed at stage ω are all possible collections of items formed at stages earlier than ω – that is, the items formed at stages 0, 1, 2, 3, etc. After stage ω, there are stages $\omega + 1$, $\omega + 2$, $\omega + 3$, etc., each of which is obtained by forming all possible collections of items formed at the preceding stage and collecting up what came before. After all of these stages, there is a stage $\omega + \omega$, i.e. stage $\omega \cdot 2$. The sets formed at this stage are obtained by forming all possible collections of items formed at previous stages – that is, the items formed at stages $\omega + 1$, $\omega + 2$, $\omega + 3$, etc. And we keep on going. So we have a stage $\omega \cdot 2 + 1$, and so on, and so forth: in general, if α is 0 or a successor ordinal, the sets formed at stage α are the sets formed at earlier stages and all arbitrary collections of these; if α is a limit ordinal, the sets formed at stage α are simply all arbitrary collections of sets formed at earlier stages.

The result of this rough description is a picture of the set-theoretic universe as a *cumulative hierarchy* divided into levels (see Figure 2.1). If we identify levels with sets and the operation of forming all possible collections of previously given objects with the powerset operation, the picture can be captured by the following characterization. Let α be an ordinal. Then, the levels of the hierarchy, **V**, are organized as follows:[1]

$$V_0 = \{x : x \text{ is an individual}\};$$

$$V_{\alpha+1} = V_\alpha \cup \mathcal{P}(V_\alpha);$$

$$V_\lambda = \bigcup_{\alpha < \lambda} V_\alpha \text{ if } \lambda \text{ is a limit ordinal.}$$

---

[1] Note that this is a definition by transfinite recursion (see Appendix 1.A). Transfinite recursion is available in the theory ZFC, to be presented below.

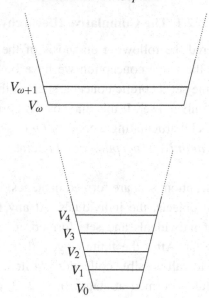

$V_{\omega+1}$

V_ω

V_4

V_3

V_2

V_1

V_0

Figure 2.1 A picture of the cumulative hierarchy.

In other words, the hierarchy is the structure obtained by starting with a collection of objects (possibly empty), taking its powerset, *its* powerset, and so on, collecting up as we proceed. The construction is indexed by the ordinals and is iterated into the transfinite. At limit ordinals, we simply collect up what came before. According to the iterative conception, there are all and only the sets that appear at one level or another of the cumulative hierarchy.

The cumulative hierarchy is due to John von Neumann (1925) and Ernst Zermelo (1930). Caution is in order, however, with regard to attributing the iterative conception to these authors. For the former simply noticed that the hierarchy is an inner model of the axioms of Zermelo set theory without the Axiom of Foundation (see Section 2.2). The latter does adopt Foundation – arguably the central axiom of the iterative conception – but in a rather tentative manner:

This last axiom, which excludes all 'circular' sets, all 'self-membered' sets in particular, and all 'rootless' sets in general, has always been satisfied in all practical applications of set theory, and, hence, does not result in an essential restriction of the theory *for the time being*. (Zermelo 1930: 403, my emphasis)

Zermelo's point seems to be that we do not *need* sets that do not appear in the cumulative hierarchy.

It is easy to see that the cumulative hierarchy picture of the set-theoretic universe endorses the set concept's indefinite extensibility: given a set a occurring at a certain level of the hierarchy, there will always be a subsequent level where the larger set obtained by performing the set-formation operations on a occurs. The picture, moreover, explains the rejection of universality: the sets occurring at each level have only members of preceding levels; thus, no set can be a member of itself, which ensures that there can be no set of all sets. For if there were, it would have itself as a member. From this, it is also easy to see why Russell's Paradox does not arise: the Russell set would be a universal set, and, as we have seen, there can be no such set on the iterative conception.

There are three key components to the account: the individuals, with which the construction begins; the ordinals, along which the construction is iterated; and the powerset operation, thanks to which new sets are built as we go from one stage to its successor. To these three components, there correspond as many questions about the structural features of the hierarchy.

Individuals

First, there is the question of what the objects we start with are: what should we take the individuals to be? We could take them to be people or fundamental particles. Or we could take them to be the integers (see Gödel *1951: 306 and Gödel 1964: 258) or, more generally, some preferred number system (see Oliver and Smiley 2006: 144–151). Or we could take the iterative process to begin with *no* individuals at all (see Figure 2.2), in which case all sets will be *pure* – where a set is pure if and only if all of its members are sets, all members of its members are sets, and so on.[2] In this case, it is no longer necessary to collect up as we proceed, and the characterization of the cumulative hierarchy takes the following simplified form:

$$V_0 = \emptyset;$$
$$V_{\alpha+1} = \mathcal{P}(V_\alpha);$$
$$V_\lambda = \bigcup_{\alpha < \lambda} V_\alpha \text{ if } \lambda \text{ is a limit ordinal.}$$

From the mathematician's perspective, starting with no individuals makes a lot of sense: mathematicians tend to be interested in structures up to isomorphism, and it is usually assumed that – no matter how complex or big a putative set of individuals might be – there will always be a corresponding set in the hierarchy of the same order type (see Reinhardt 1974: 190).

Thus, from the point of view of the mathematician, the restriction to pure sets, while making things technically smoother, is not a real restriction, since it has no substantial bearing on the richness of the structure under consideration. However,

---

[2] In technical terms, we say that a set is pure iff it does not contain individuals and neither do the members of its *transitive closure*. The transitive closure of a set *a* is the smallest transitive set which contains *a*.

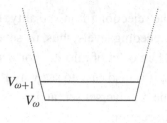

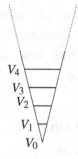

Figure 2.2 A picture of the pure cumulative hierarchy.

the fact that in certain contexts we restrict attention to pure sets should not be taken
as indicating that there are no sets with individuals: were this to be the case, the
applicability of set theory would have to be given up, or at least be much harder to
ensure (see Potter 2004: 76–77). Rather, the point is simply that for certain purposes
we can ignore sets with individuals and confine attention to pure sets.[3] Since I shall
mostly be concerned with the justification of the relevant set theories, rather than
with their development, I will adopt a mixed policy: for the sake of simplicity, I
will often confine attention to pure set theories; however, I shall occasionally revert
to impure ones when required.

   Also, note that taking levels to be sets implies that there is a set of all individuals.
The usual argument for this is that the iterative picture of the set-theoretic universe
makes it hard to see how the individuals could fail to form a set. As David Lewis
puts it:

[T]he obstacle to sethood is that the members of [a plurality which does not form a set] are
not yet all present at any rank of the iterative hierarchy. But all the individuals, no matter
how many there may be, get in already on the ground floor. So, after all, we have no notion
what could stop any class of individuals [...] from comprising a set. (Lewis 1991: 104)

---

[3] Hence, the best way of making sense of mathematicians' claims that individuals are not needed for mathematical
purposes – criticized, for instance, by Oliver and Smiley (2006: 145–146) – is to construct 'mathematical
purposes' as the purposes of the pure mathematician, whose only interest is in features of structures up to
isomorphism.

This, however, is not uncontroversial, and Zermelo (1930) himself did not assume the individuals to form a set. For discussion, see Parsons 1977: 504, fn. 4, Menzel 2014: §2 and Uzquiano 2015a: 287–288, which also provides a formal characterization of the cumulative hierarchy which does not require the individuals to form a set.

Height

The second and third questions about the structural features of the cumulative hierarchy do have a substantial bearing on its richness, and are therefore relevant even from the mathematician's point of view. To use a slogan, they concern the *height* and *width* of the hierarchy.

Let us consider the issue of the height of the hierarchy first. We are told that the cumulative process of construction is indexed by the ordinals, but how far does this process go? An initial and frequently given answer is that the process should be iterated *as far as possible*. But again, this does not tell us very much until the idea of iteration 'as far as possible' is developed to some extent.

One possible way of spelling out the idea of iteration 'as far as possible' consists in saying that the process of construction should go as far as the ordinals go. That is to say, the construction process is to be iterated along *all* the ordinals.

This suggestion goes some way towards answering the second question and delivers what we shall later refer to as the Axiom of Ordinals (see Section 3.6). However, the suggestion also does not tell us very much: we are told that the construction process goes as far as the ordinals, but how far do the *ordinals* go? The picture of the set-theoretic universe as a cumulative hierarchy divided into levels explicitly refrains from telling us much about this: it leaves it open what counts as an arbitrary ordinal.

To answer *this* question, a variety of principles have been invoked. On the one hand, we have principles telling us, effectively, that the hierarchy goes at least as far as a certain ordinal. These include the Axiom of Infinity

$$\exists y(\emptyset \in y \land \forall x(x \in y \rightarrow x \cup \{x\} \in y)),$$

which states that there is an infinite set,[4] and the so-called *large cardinal axioms*, which assert the existence of certain larger sets such as inaccessible, Mahlo, weakly

---

[4] This is the version of the Axiom of Infinity standardly used nowadays. But there are other versions of the axiom which are not first-order equivalent to it, such as Zermelo's (1908). This asserts the existence of a set which contains the empty set and the singleton of every set it contains, that is: $\exists y(\emptyset \in y \land \forall x(x \in y \rightarrow \{x\} \in y))$. However, the differences between these two axioms collapse in the presence of axiom (ρ), which we shall discuss below and take as part of our base theory. See Uzquiano 1999 for details.

compact, ω-Erdös and measurable cardinals. On the other hand, we have the so-called *reflection principles*, which tell us that the hierarchy has certain absoluteness properties and should, *as a result*, go at least as far as a certain ordinal.[5]

Width

We shall have occasion to discuss reflection principles in Section 3.6. For the moment, let us turn to the question of the width of the hierarchy, i.e. of how rich each level of the hierarchy is. The characterization of the levels of the hierarchy given above guarantees that each level $V_{\alpha+1}$ of the hierarchy is the union of the immediately preceding level V_{α} and of its powerset. This might seem to settle once and for all the question of how rich each level of the hierarchy is. In a way, this is true: besides the items occurring at the preceding level, $V_{\alpha+1}$ will contain *all subsets* of V_{α}.

But at the formal level, this does not tell us very much until we are told what should count as a subset of a given set. This is done by the Axiom of Separation, which specifies what subsets of a given set there are and gives the Powerset Axiom – asserting that for every set A, its powerset $\mathcal{P}(A)$ exists – its strength.[6] Using the language of second-order logic, we can formulate the Axiom of Separation as follows:

$$\forall F \forall z \exists y \forall x (x \in y \leftrightarrow x \in z \wedge Fx).$$

In first-order theories, however, Separation is a *schema*, with an instance for each property expressible in the language of set theory. That is:

$$\forall z \exists y \forall x (x \in y \leftrightarrow x \in z \wedge \varphi(x)),$$

where $\varphi(x)$ is, again, any formula in $\mathcal{L}_{\in}$ in which x is free and which contains no free occurrences of y.

Recall, though, that the guiding idea of the informal account of the iterative conception is that at every stage of the formation process we are supposed to form all possible collections of items formed at earlier stages. Putting this in terms of the

---

[5] In set theory, a formula φ is said to be *absolute* between models $\mathfrak{A}$ and $\mathfrak{B}$ of set theory just in case $\mathfrak{A} \models \varphi \leftrightarrow \mathfrak{B} \models \varphi$. To say that the hierarchy has certain absoluteness properties is to say that we can find interesting models $\mathfrak{A}$ and classes of formulae Φ such that every φ in Φ is absolute between $\mathfrak{A}$ and V. This means that to ascertain the truth of the formulae in φ it suffices to ascertain whether they are true in $\mathfrak{A}$. See Hallett 1984: 206–207 and Jech 2003: 163, 185 for more on the notion of absoluteness.

[6] As Gödel puts it: 'The axiom of subset formation comes before the axiom of power set. We can form the power set of a set, because we understand the selection process (of singling out any subset from the given set) intuitively, not blindly' (the remark is reported in Wang 1996: 259). The point is also made by Clark (1993: 242–243).

notion of a subset, we want to collect together *all* subsets of sets already formed, not only those defined by properties expressible in the language of first-order set theory. The third structural question is then: what counts as an arbitrary subset of a given set? The answer to this question will affect the richness of the levels of the cumulative hierarchy.

To answer the third structural question whilst sticking to first-order logic, two strategies seem available. The first and more direct strategy attempts to answer the question by making the notion of an arbitrary subset more precise. One famous instance of this strategy identifies an arbitrary subset with a *describable* subset – more specifically, with a subset describable by means of a formula of the language of set theory. This leads to a refinement of the earlier account of the cumulative hierarchy. As before, let α be an ordinal. Then the levels of the *constructible hierarchy*, **L**, are organized as follows:

$$L_0 = \{x : x \text{ is an individual}\};$$
$$L_{\alpha+1} = L_\alpha \cup \text{def}(L_\alpha);$$
$$L_\lambda = \bigcup_{\alpha < \lambda} L_\alpha \text{ if } \lambda \text{ is a limit ordinal,}$$

where $\text{def}(L_\alpha)$ is the set of all sets that are definable by means of quantifiers which are restricted to range only over L_α. The *Axiom of Constructibility* is then the assertion that all sets occur at one level or another of the constructible hierarchy, that is that all sets are constructible (**V = L**).

The second strategy contemplates the possibility of adding further axioms to set theory to ensure that at each stage more and more subsets of a given set – beyond those whose existence is guaranteed by the Axiom Schema of Separation – are collected together. Consider, for instance, the Axiom of Choice (**AC**), one version of which states that for every set a of disjoint non-empty sets there is a set c, called a *choice set* for a, which contains exactly one member of each of the members of a. It has been argued that **AC** is justified on the iterative conception on the grounds that it guarantees that at each level more subsets of a given set, namely the choice sets, are formed than those specified by the Separation Schema. In a nutshell: if each level of the hierarchy contains *all* subsets of the previous levels, then it will *a fortiori* contain the choice sets.[7] Thus, Georg Kreisel (1980: 192) writes that '[f]or the fat (or "full") hierarchy, the axiom of choice is quite evident'.

---

[7] Note that AC is also a consequence of the Axiom of Constructibility. But it is widely thought that, in the context of **L**, AC does not hold because it ensures that at each level more subsets are formed. Rather, it holds in **L** because if all sets are constructible, an artificially simple well-ordering of the universe is available (see Maddy 1988: 497). And the existence of such a well-ordering implies AC.

The argument is at first sight convincing. But it is not without problems. For, as Boolos (1971: 28–29; 1989: 96–97) pointed out, the argument seems just to *assume* that the choice sets are among the subsets of a given set: the cumulative hierarchy picture of the set-theoretic universe assures us that at each level we can collect together all collections of items formed at earlier stages, but not that the choice sets *are* formed.

The argument might have a better fate if understood not as purporting to establish that the cumulative hierarchy picture forces AC upon us, but that AC is the hallmark of a *combinatorial* conception of collection (see Section 1.8): if the existence of a set does not depend on the existence of a condition satisfied by all the members or of a rule for selecting them, then nothing seems to stand in the way of the choice sets being formed, or so the thought goes. Indeed, this appears to be the version of the argument given by, for instance, Ramsey (1925: 220–221),[8] Bernays (1935: 260)[9] and Paseau (2007: 34–35).[10] And even if this combinatorial conception cannot quite be *subsumed* under the iterative conception, it does seem to harmonize quite well with it: the iteration process is naturally understood in a combinatorial manner.[11]

2.2 Iterative Set Theories

Hence, the picture of the set-theoretic universe as a cumulative hierarchy divided into levels leaves it open what counts as an arbitrary ordinal and – at the formal level at least – as an arbitrary subset. Nonetheless, it is usually agreed that the picture suffices to sanction (at least some of) the axioms of standard set theory –

---

[8] 'If by "class" we mean, as I do, any set of things homogeneous in type not necessarily definable by a function which is not merely a function in extension, the Multiplicative Axiom seems to me the most evident tautology'.

[9] 'The axiom of choice is an immediate application of the quasi-combinatorial concepts in question [i.e. the concepts of set, function, and sequence]'.

[10] 'If one thinks that *any* arbitrary combination of sets below some given stage constitutes a property, then a generalisation of *Spec* to cover all possible properties whatsoever – as opposed to those expressible in some formal language, as in Boolos's initial presentation – expresses the intuitive thought that at any given stage all the possible sets available for formation are indeed formed. As it is usually conceived, and as Boolos himself conceives it, the iterative conception incudes this combinatorial idea. And combinatorialism straightforwardly implies a Choice axiom'.

[11] Another strategy consists in arguing that AC is true on the iterative conception because it can be derived from some suitable second-order logical principle. For instance, Pollard (1988, 1992) suggests that the axiom should be regarded as true on the iterative conception because it follows from a principle of plural choice, and, similarly, writers such as Tait (1998) and Burgess (2004) have suggested adding a choice principle to the background second-order logic. Indeed, such a principle was included by Zermelo (1930) in the paper in which he described the cumulative hierarchy. What Boolos would probably have said in reply to this is that deciding whether to accept those principles is no less hard than deciding whether to regard AC as true. Compare Boolos 1971: 28: 'The difficulty with the axiom of choice is that the decision whether to regard the rough description as implying a principle about sets and stages from which the axiom could be derived is as difficult a decision, because essentially the same decision, as the decision whether to accept the axiom'.

by which I mean ZFC, i.e. first-order Zermelo-Fraenkel set theory (ZF) with AC. ZFC's quantifiers only range over sets, but one can also formulate a version of it which admits individuals, namely ZFCU. (In general, when a theory of set admits of individuals, this is indicated by adding a 'U' – from '*Urelemente*' – to the end of its name: so we have ZFCU, ZFU, and so on.) ZFC has the resources to reconstruct most if not all of ordinary mathematics, and hence to act as a foundation for mathematics. Moreover, it can develop a substantial theory of the infinite ordinal and cardinal numbers, including, notably, proving Cantor's Theorem.

Let us first explore the axioms of a subsystem of ZFC, namely Zermelo set theory Z. In Z, we have two sorts of axioms of extent. On the one hand, we have axioms for constructing sets out of given ones. These include Separation, Powerset and two other axioms: the Axiom of Union, asserting that for every set there exists its union,[12] and the Axiom of Unordered Pairs, which states that given any set a and b, there exists a set containing exactly a and b. On the other hand, we have two axioms making an outright existential assertion, namely the Axiom of Empty Set, stating that there is an empty set, and the Axiom of Infinity. The former is often omitted because, in the presence of Infinity, it is an easy consequence of Separation.

We then have two axioms of nature, the Axiom of Extensionality and the Axiom of Foundation. The reason for including the Axiom of Extensionality is usually taken to be that, as we saw in Chapter 1, it serves to distinguish sets from other entities. On these grounds, Boolos (1971, 1989) famously argued that Extensionality does not follow from the iterative conception. In effect, however, Boolos's point seems to be that, strictly speaking, we should perhaps say that Extensionality is true on the iterative conception *and the concept of set*. This point, however, seems to apply to other conceptions of set as well, as Alexander Paseau (2007: 44–46) has, to my mind, convincingly argued. For this reason, and for ease of exposition, I will therefore often omit the qualification in the remainder of this book.

In its usual formulation, the Axiom of Foundation asserts that every non-empty set a contains an ∈-*minimal* element, i.e. an element disjoint from a. That is to say, the axiom states that every set is well-founded. The axiom is *intended* to guarantee that if you pick a member of a, and then a member of a member of a, and then a member of a member of a member of a, you eventually end up with a set which has no members (or with an individual). To use a standard expression, the axiom is intended to guarantee that there are no infinite descending ∈-chains, that is chains of the form

$$a_0 \ni a_1 \ni a_2 \ni \ldots$$

---

[12] The union of a set a is the set whose members are the members of the members of a.

Hence, the axiom is intended to rule out the existence of, for instance, any set a such that $a \in a$ and of closed membership chains such as

$$a_0 \in a_1 \in a_2 \in a_0.$$

But what reason is there for regarding the axioms of Z_i (Z minus Extensionality) as true on the iterative conception? One reason that has been offered is that they follow from a theory of stages ST formalizing (part of the pure version of) the rough description of the iterative conception.[13]

ST is cast in a two-sorted first-order language with variables $x, y, z \ldots$ for sets and variables $r, s, t \ldots$ for stages. The language also includes three two-place predicates: a stage-stage predicate $<$ ('is earlier than'), a set-stage predicate F ('is formed at') and the membership predicate. We abbreviate $\exists t (t < s \wedge yFt)$ ('y is formed before s') as yBs. The axioms of ST are the following five axioms

$$Tra \quad \forall t \forall s \forall r (t < s \wedge s < r \rightarrow t < r),$$
$$Net \quad \forall t \forall s \exists r (t < r \wedge s < r),$$
$$Inf \quad \exists r (\exists t\, t < r \wedge \forall t (t < r \rightarrow \exists s (t < s \wedge s < r)),$$
$$All \quad \forall x \exists s\, xFs,$$
$$When \quad \forall x \forall s (xFs \leftrightarrow \forall y (y \in x \rightarrow yBs)),$$

plus all instances of

$$Spec \quad \exists s \forall y (A(y) \rightarrow yBs) \rightarrow \exists x \forall y (y \in x \leftrightarrow A(y)),$$

where $A(y)$ is a formula of the language of ST which does not contain x free.

We then have the following (see Boolos 1989: 103–104):

Theorem 2.2.1 (Shoenfield-Boolos). ST *implies the axioms of* Z_i.

Insofar as the stage theory captures part of the content of the iterative conception, the fact that the axioms of Z_i follow from those of the stage theory shows that they are justified on the iterative conception.

However, the *origins* of Z do not lie with the iterative conception. Instead, when Zermelo (1908) laid down the first axiomatization of set theory, of which Z is a descendant, his motivation seems to have been to clarify the assumptions involved in his 1904 proof of the Well-Ordering Theorem, stating that every set can be well-ordered. Thus, Zermelo's original axiomatization did not include the Axiom of Foundation but included the Axiom of Choice, of which essential use is made in his proof. Indeed, Zermelo (1908) explicitly refrains from claiming that his axiomatization captures an intuitive conception of set (see Section 5.1 for details).

---

[13] ST is presented by Boolos (1989), building on Boolos 1971 and Shoenfield 1977.

Table 2.1 *The axioms of* ZFC

Name	Formula and Verbal Description
EXTENSIONALITY	$\forall x \forall y (\forall z (z \in x \leftrightarrow z \in y) \rightarrow x = y)$. No two distinct sets have the same members.
FOUNDATION	$\forall x (x \neq \emptyset \rightarrow \exists y (y \in x \wedge \neg \exists z (z \in x \wedge z \in y)))$. Any non-empty set contains a set with which it has no members in common.
EMPTY SET	$\exists y \forall x (x \in y \leftrightarrow x \neq x)$. There exists a set with no members.
INFINITY	$\exists x (\emptyset \in x \wedge \forall y (y \in x \rightarrow y \cup \{y\} \in x))$. Say that a set A is *inductive* iff the empty set belongs to A and, for any x, if x belongs to A, then $\{x\}$ belongs to A. Then, there is an inductive set.
UNORDERED PAIRS	$\forall z \forall w \exists x \forall y (y \in x \leftrightarrow (y = z \vee y = w))$. For any set z and w there is a set whose sole members are z and w.
UNION	$\forall z \exists x \forall y (y \in x \leftrightarrow \exists w (y \in w \wedge w \in z))$. For any set z, there is a set whose members are precisely the members of members of z.
POWERSET	$\forall z \exists x \forall y (y \in x \leftrightarrow y \subseteq z)$. For any set z, there is a set whose members are precisely the subsets of z.
SEPARATION	$\forall z \exists x \forall y (y \in x \leftrightarrow (y \in z \wedge \varphi(y)))$. For any set z and any condition φ, there is a set whose sole members are the members of z which satisfy φ.
REPLACEMENT	$\forall v \exists! w \varphi(v, w) \rightarrow \forall u \exists y \forall x (x \in y \leftrightarrow \exists v (v \in u \wedge \varphi(v, x)))$. The image of a set in a first-order definable function is a set.
CHOICE	$\forall u ((\forall x (x \in u \rightarrow \exists u (u \in x)) \wedge \forall x \forall y (x \in u \wedge y \in u \wedge x \neq y \rightarrow \neg \exists w (w \in x \wedge w \in y)) \rightarrow \exists z \forall x (x \in u \rightarrow \exists! w (w \in x \wedge w \in z))$. For any set u of pairwise disjoint non-empty sets, there exists a set z that contains exactly one element in common with each of the sets in u.

ZF adds to Z the Axiom Schema of Replacement. This is an axiom of extent of the kind that makes a conditional existential assertion. In particular, it says that the image of a set in a function is a set. The issue whether Replacement is justified on the iterative conception is a vexed one, and we shall discuss it in Section 3.6.

ZFC – whose axioms are summarized in Table 2.1 – is obtained by adding the Axiom of Choice to ZF. Choice does not fit neatly into the division of axioms into those of nature and those of extent,[14] and as noted in the previous section, it is disputed whether it is sanctioned by the iterative conception.

---

[14] It is perhaps worth noting that this fits well with the view that AC should be accepted because it follows from some suitable second-order logical principle.

Because of their controversial status on the iterative conception, we shall not include Replacement and Choice in the base theory used to discuss the conception. Instead, we shall often be interested in a theory slightly stronger than Z, which we shall call Z^+ (see Uzquiano 1999). This theory is obtained from Z by replacing the Axiom of Foundation with the following stronger axiom of nature, which asserts that every set is the subset of some level of the hierarchy:

$$\forall x \exists \alpha \exists y (y = V_\alpha \wedge x \subseteq y).^{15} \tag{ρ}$$

The theory Z^+ is equivalent to another theory based on the cumulative hierarchy, which I shall refer to as SP (from Scott-Potter). This is a modification of the pure version of the theory presented by Michael Potter (2004), and builds on Dana Scott's (1974) axiomatization of set theory.

Scott's axiomatization uses a two-sorted language, with variables $x, y, \ldots$, ranging over sets, and variables $V, V', \ldots$, ranging over levels ordered by membership. (Scott's term is 'partial universes'.) The distinguishing axioms of Scott's theory are the Axiom of Accumulation – stating that the members of a level are the members or subsets of earlier levels – and the Axiom of Restriction – asserting that every set is a subset of some level. Effectively, then, the Axiom of Restriction is the same as axiom (ρ). Scott then showed that the axioms of Accumulation and Restriction – together with the axioms of Extensionality and Separation and an assumption to the effect that every set is a member of some level – imply all the axioms of Z minus Infinity.

SP differs from Scott's axiomatization in that its language is one-sorted and the notion of a level is not taken as a primitive but is explicitly defined in terms of membership.[16] Moreover, Scott's Axiom of Accumulation becomes a consequence of the definition of levels in terms of membership. Finally, we have an Axiom of Creation, which states that there is no highest level, and an Axiom of Infinity, asserting the existence of a limit level.[17]

It is then easy to see that the following is the case:

---

[15] In this case, rather than being defined by transfinite recursion, the V_αs are defined by synthetic means as follows:

$$x = V_\alpha \Leftrightarrow \exists f (\text{Dom}(f) = \alpha + 1 \wedge \forall \beta \leq \alpha \forall y (y \in f(\beta) \leftrightarrow$$
$$\exists \lambda < \beta (y \subseteq f(\lambda))) \wedge f(\alpha) = x).$$

[16] The details are as follows. We first characterize the notion of accumulation in terms of membership by defining the *accumulation* of a set a, acc(a), as the set whose members are all the members and subsets of all the members of a. (When individuals are admitted, the accumulation of a set will also contain all the individuals.) Then, we say that a set h is a *history* iff $\forall v \in h(v = \text{acc}(h \cap v))$. Finally, we define a *level* as the accumulation of a history.

[17] SP is a modification of the system used in Potter 2004 in that the latter does not include the Axiom of Restriction and is officially a theory of *collections*. In Potter's system, sets are then defined as subcollections of

Proposition 2.2.2. Z^+ *is equivalent to* SP.

Sketch of proof. The left-to-right direction follows from the fact that in Z^+ we can show that the V_αs are all the levels, which enables us to derive the Axiom of Creation and SP's Axiom of Infinity (see Potter 2004: 296). The right-to-left direction follows from Scott's (1974) proof that the axioms of Z are theorems of his set theory, and from the fact that the Axiom of Restriction is essentially the same as axiom (ρ). $\square$

An advantage of Z^+ over Z is that, in the absence of Replacement, the standard formulation of Foundation fails to enforce the cumulative hierarchy structure on the set-theoretic universe. This is a reflection of the fact that the standard proof that Foundation implies that there are no infinite descending $\in$-chains makes use of the Replacement Schema, and is better appreciated by looking at what happens to the two theories when we switch from the first- to the second-order case.

One of the reasons for being interested in second-order theories is that it is sometimes possible to prove that they are *categorical* – that is, that all their models are isomorphic.[18] As an example, consider PA_2, the second-order theory of arithmetic obtained by replacing Peano Arithmetic (PA)'s Axiom Schema of Induction with the corresponding second-order axiom.[19] This theory categorically characterizes the natural numbers as the smallest structure closed under a one-to-one successor operation and containing an element which is not the successor of any element.

In the case of iterative set theory, the best we can hope for is *quasi-categoricity*:[20] given two models $\mathfrak{A}$ and $\mathfrak{B}$, either there is an isomorphism from the pure part of $\mathfrak{A}$ to an initial segment of the pure part of $\mathfrak{B}$, or viceversa. Using the terminology

some level of the hierarchy. Here, we are following Scott in keeping the Axiom of Restriction and formulating our theory as a theory of *sets*.

[18] Although models are typically taken to be sets, set-theorists sometimes speak of 'class models', models in which the domain is not a set but a class. Usually, however, this talk of classes is eliminable, since the classes in question are *definable* classes, which must not be thought of as new objects, but as abbreviations for expressions not involving them (see Jech 2003: 5 for details). In Section 3.3 we shall consider the case where it looks as though we are dealing with models in which the domain is not a set but it will not do to appeal to definable class models.

[19] Peano Arithmetic is the first-order theory consisting of an axiom stating that 0 is not the successor of any number, an axiom stating that every number has a successor, an axiom stating that the successor operation is one-to-one, the recursion equations for addition and multiplication, and the Axiom Schema of Induction, which states that mathematical induction holds when applied to a formula expressible in the first-order language of arithmetic. Note that the extra strength afforded by the second-order Induction Axiom also means that addition and multiplication are definable in PA_2 and thus their recursion equations no longer need to be laid down as axioms.

[20] McGee (1997) has argued that we can indeed prove full categoricity for set theory. See Incurvati 2016 for a critical discussion.

of the previous section, for pure second-order set theories it is often possible to prove that their models can only differ in height. It turns out, however, that Z_2 – the theory one obtains by replacing Z's Axiom of Separation with the corresponding second-order Axiom – is not quasi-categorical, the reason being that it admits of non-well-founded models (see Uzquiano 1999). Such models, on the other hand, are ruled out by axiom (ρ), which enforces the cumulative hierarchy structure upon the set-theoretic universe. Unsurprisingly, then Z_2^+ is quasi-categorical.[21]

Axiom (ρ) also enables us to overcome a number of other shortcomings of Z and its second-order counterpart, namely their failure to: guarantee the existence of sets that appear at V_ω (Uzquiano 1999: 290–294); cope satisfyingly with transfinite recursion (see Mathias 2001); prove that every set has a transitive closure. The latter, in particular, makes the existence of non-well-founded models of Z_2 much less surprising. For the fact that every set has a transitive closure (which follows from Replacement) is one of the facts used to prove that Foundation implies the second-order version of the principle of transfinite induction on $\in$ in the context of the axioms of ZF_2 – where ZF_2 is, as usual, the theory one obtains if one replaces ZF's Replacement Schema with the corresponding second-order axiom.[22] And it is the principle of second-order $\in$-induction which enables us to prove that ZF_2 is only satisfiable in models in which the membership relation is well-founded,[23] a result which was essentially first proved by Zermelo (1930).[24]

For these reasons, and because the iterative conception arguably justifies the thought that every set is a subset of some level, we will take Z^+ as our base theory for the iterative conception in the remainder of this book.[25]

---

[21] Which is not to deny that, as we said in the previous section, besides the issue of how high the hierarchy is, there is also the issue of how wide it is. For there is a sense in which the richness of the powerset operation is not settled simply by switching to the second-order level, since issues analogous to the question of what should count as a subset of a given set seem to arise for second-order logic itself. I discuss these matters in Incurvati 2008.

[22] It is also customary not to include second-order Separation in the axiomatization, since that is already a consequence of second-order Replacement. Here are some more details concerning the principles under consideration and how they are related. The first-order version of $\in$-induction is: $\forall x(\forall y \in x\ \varphi(y) \to \varphi(x)) \to \forall x\varphi(x)$. This is a contrapositive of the schema: $\exists x\varphi(x) \to \exists x(\varphi(x) \wedge \forall y \in x\neg\varphi(y))$. In the presence of the other axioms of ZF, all other instances of this schema follow from the Axiom of Foundation. A similar reasoning is used to show that the second-order version of the principle of $\in$-induction follows from the Axiom of Foundation in the presence of the axioms of ZF_2.

[23] Indeed, Uzquiano (1999: 297 and 299–301) shows that even if we replace Z_2's Axiom of Infinity with an axiom capable of guaranteeing the existence of sets that appear at V_ω, and add to the resulting theory an axiom stating that every set has a transitive closure, we still will not end up with a theory adequate to characterize the initial segments of the cumulative hierarchy which are indexed by a limit ordinal $\lambda > \omega$. On the other hand, such segments of the hierarchy are adequately characterized by Z_2^+.

[24] Zermelo's original result concerns a theory without Infinity, which can also be satisfied in V_ω.

[25] Uzquiano (1999: 301) claims that Z^+ is not 'a natural extension of the Zermelo axioms'. However, he does not offer any argument for this conclusion. And in fact, to the extent that it sanctions the axioms of Z, the idea that

2.3 Priority of Construction

According to the iterative conception, then, sets can be arranged in a cumulative hierarchy divided into levels. This conception sanctions (at least) most of the axioms of standard set theory and provides a convincing explanation of the paradoxes; but is it *correct*? In other words, what reasons do we have for thinking that sets and levels are as the iterative conception says they are? The standard answer to this question is what I shall refer to as the *substantial* approach.

The substantial approach attempts to defend the view that all sets are those in the hierarchy by arguing that this follows from general considerations about the nature of sets. Usually, this is done by claiming that 'there is a fundamental relation of presupposition, priority or [...] *dependence* between collections' (Potter 2004: 36). Sometimes, the iterative conception is described simply as the conception of set one ends up with if one takes sets as bearing such a relation of priority or dependence between them (see, e.g., Potter 2004: 36). Now, the crucial question for the substantial approach is to say what this relation amounts to. We can distinguish two answers to this question: a *constructivist* one and a *platonist* one.

According to the constructivist, talk of set formation in the informal description of the iterative conception has to be taken literally: at each stage of the construction process, sets *really are* formed by collecting together any objects that are available – where some objects are available just in case they are independently given to us (presumably in our thought) or have been formed out of objects already so given (and are thereby, it is assumed, also given in our thought).

It follows that the relation of priority between a set and its members is, for the constructivist, just the relation of priority in the order of construction. Whilst providing the defender of the iterative conception with a clear answer to the question of what the relation of priority amounts to, however, the constructivist approach also implies that there are severe limitations to the strength of the theory which the iterative conception ends up sanctioning.

For one thing, it is not clear whether the idea that sets are formed by collecting together any given objects can be extended to the case where the size of the given objects is uncountable (see Potter 2004: 37). For another, it seems part of the constructivist doctrine that, at any point in the construction process, we can only construct sets specifiable by reference to sets already constructed.[26] This, however, seems to sanction only a predicative version of Z's Separation Schema, which

---

sets are the objects that occur at some level of the cumulative hierarchy seems to sanction axiom (ρ) too. In any case, Proposition 2.2.2 tells us that Z^+ is equivalent to a natural axiomatization of the iterative conception.

[26] One might try to resist this, but the challenge then becomes to articulate an account of the iterative process such that it can transcend definability without jeopardizing its constructivist credentials.

drastically cuts down the strength of another of the theory's axioms – the Axiom of Powerset. For, as pointed out above, the Powerset Axiom gets its strength only in conjunction with the Separation Schema, which specifies, by means of a formula φ, which subsets of a given set there are. And in Z this formula can contain quantifiers ranging over anywhere in the set-theoretic universe: the standard version of the Separation Schema is *impredicative*. To put it another way, in standard set theory the size of the powerset of a set a depends not only on the size of a, but also on the richness of the entire universe.[27] This seems incompatible with any constructivist account according to which the set-theoretic universe is conceived of as being created in stages by using the powerset operation itself.

The problem is not the restrictions per se: a constructivist might well be happy to reject any part of classical mathematics which entails the existence of uncountable sets and makes use of impredicative specifications. Rather, the problem is that the constructivist approach is often invoked to argue that the iterative conception is correct *and* sanctions the axioms of standard set theory, or at least those of Z. Thus, for instance, Hao Wang writes that, on the basis of his explanations (largely based on a constructivist approach), 'we are able to see that the ordinary axioms of set theory (commonly referred to as ZF or ZFC) are true for the concept of set' (1974: 184), where, he notes, 'we do not concern ourselves over how a set is defined, e.g. whether by an impredicative definition' (1974: 183). On the other hand, he admits that 'if we adopt a constructive approach, then we do have a problem in allowing unlimited quantifiers to define other sets' (1974: 209).

Shapiro disagrees with this conclusion. Building on some remarks by Wang himself,[28] he attempts to overcome this problem in an ingenious way. According to him, the foregoing considerations do not

preclude the possibility of impredicative definition. It turns out that the legitimacy of impredicative definition is a substantial set-theoretic principle. In the present framework, the ideal constructor, and thus a human counterpart, can define sets by reference to any fixed rank V_α. That is, a definition is legitimate if its variables are restricted to a fixed rank. This allows some impredicative definitions, and it may allow all. If the underlying language is first-order, then there is a reflection theorem to the effect that if a formula is

---

[27] Set-theorists put the matter by saying that the formula $y = \mathcal{P}(x)$ is not absolute for transitive models. If φ is absolute for transitive models, to ascertain whether the property expressed by φ holds of a set, it is enough to ascertain whether φ holds of that set in any transitive model (and similarly for the case of relations). This cannot be done in the case of $y = \mathcal{P}(x)$, since y may be the set of all subsets of x in a certain transitive model and yet not be its powerset in the cumulative hierarchy: we can never rule out that, when we consider the whole set-theoretic universe, subsets can become definable that were not definable before, thereby affecting the truth-value of $y = \mathcal{P}(x)$.

[28] 'We may also appeal to the reflection principles to argue that the unbounded quantifiers are not really unbounded' (Wang 1974: 209).

true at all, then it is true at some sufficiently large rank. Thus, the ideal constructor never has to reach 'arbitrarily high' in the hierarchy in order to characterize a set. This sanctions each instance of the axiom of separation and allows full impredicative definition. (Shapiro 1997: 202)

We shall come back to the Reflection Theorem Shapiro is referring to in Section 3.6. For the moment, it suffices to notice that the theorem is not provable in Z^+ but becomes provable when we add Replacement to it. As already pointed out, this is a schema, which delivers the Reflection Theorem about *all* formulae of set theory only if it is itself understood impredicatively. Hence, if the constructivist is to avail herself of the required Reflection Theorem, she has to offer a justification for the *impredicative* version of Replacement. And it is far from clear why this should be any easier than the justification for the impredicative version of Separation. Indeed, if impredicative Replacement could be justified on constructivist grounds, then there would be a much shorter route to impredicative Separation, since, as already pointed out for the second-order case (see Section 2.2, fn. 22), the latter is an immediate consequence of the former (and is therefore often omitted in axiomatizations of ZF).

2.4 Metaphysical Dependence

Perhaps because of these sorts of difficulties, constructivist accounts of the hierarchy do not seem to be very popular nowadays, especially among philosophers. A different version of the substantial approach to be found in the literature is the platonist one (see Potter 2004; Linnebo 2008). On this account, sets are not literally formed at each stage. Rather, talk of set formation has to be taken at best as a metaphor, and this is usually signalled by the presence of scare quotes. This also seems to be the approach usually followed by more philosophically minded settheorists. Let us give a couple of examples (the first from a mathematician, the second from a philosopher):

Now surely, before we can form a collection of objects, those objects must first be 'available' to us! [...] Before we can build sets of sets of objects, we must have the sets of objects out of which to build those sets. The crucial word here, of course, is 'build'. Naturally, we are not thinking of actually building sets in any constructive sense. But our set theory should certainly reflect this idea. (Devlin 1993: 35–36)

According to the prevailing iterative conception, sets are 'formed from' their elements. The relation between a set and its elements is thus asymmetric, because the elements must be 'available' before the set can be formed, whereas the set need not be, and indeed cannot be, 'available' before its elements are formed. (Linnebo 2008: 72)

At this point, it is not rare to observe a bifurcation in the attitudes of writers. Some, whilst stressing that all we are dealing with is a metaphor, do not feel the need to explain what it is a metaphor for. For instance, Devlin (1993: 36) goes on to say that 'putting these vague considerations into a more precise setting, we see that set theory is essentially hierarchical in nature'. The problem with this attitude, however, is that, until the metaphor is explicated, it is not entirely clear why the metaphor should provide us with reasons to believe that the set-theoretic universe really is a cumulative hierarchy divided into levels. It seems that all we have here is a *heuristics*, and not a justification, for some of the axioms of standard set theory.

Other writers, on the other hand, do replace (or even dispense altogether with) metaphorical talk and claim that the relation of priority between a set and its members is a relation of *metaphysical dependence*. The idea, to borrow the standard example, is that whilst the singleton of Socrates metaphysically depends on Socrates, there is no dependence in the reverse direction. But what does it mean to say that a set metaphysically depends on its members?

It is tempting to characterize the dependence relation in terms of necessity, and say that x depends on y if and only if, necessarily, x exists only if y exists. Yet this will not serve the dependency theorist's purposes. For although she accepts that necessarily the singleton of Socrates exists only if Socrates does, she also seems to be committed to the converse: necessarily, Socrates exists only if his singleton does (see, e.g., Fine 1995: 271). For this and similar reasons, Potter (2004: 39–40) concludes that 'priority is a modality distinct from that of time or necessity, a modality arising in some way out of the manner in which a collection is constituted from its members'. The idea, therefore, seems to be that the dependence relation of a set upon its members is primitive and unanalyzable, although this, the thought goes, does not prevent us from determining some of its structural properties.

Linnebo, on the other hand, while admitting that '[t]he only feature of the notion which emerges as reasonably clear is that the dependence in question is supposed to be a matter of how one object depends on the other *for its identity*' (2008: 74), writes that 'the intuitive notion of dependence which [he has] relied on needs to be analysed' (2008: 77). To this end, he appeals to the accounts offered by Kit Fine (1994; 1995) and Lowe (2003; 2005), who, he claims, have offered '[t]wo of the most sophisticated analyses' of that notion (2008: 77).

Now, the accounts offered by Fine and Lowe do provide necessary and sufficient conditions for an object x depending on an object y. To this extent, they represent analyses of the notion of dependence. So, according to Fine, x *ontologically depends* on y if and only if y is a constituent of some essential property of x. Hence, to say that the singleton of Socrates depends on Socrates but not vice versa is to say that the property of having Socrates as an element is essential to the singleton of Socrates but the property of being an element of the singleton of Socrates is not essential to Socrates. For Lowe, on the other hand, *x depends for its identity* on *y* if and only if there is a function *f* such that it is essential to *x* that $x = f(y)$. Thus, the dependence of the singleton of Socrates upon Socrates himself is given by the fact that whilst it is essential to the singleton of Socrates that it is the value of the function $x \mapsto \{x\}$ applied to Socrates, it is not essential to Socrates that he is the value of the function $\{x\} \mapsto x$ applied to the singleton of Socrates.

But how informative are these analyses? The crucial notions in the *analysans* are, respectively, that of an essential property and that of something being essential to an object. So it is these that we have to examine to evaluate the informativeness of the proposed analyses.

Linnebo (2008: 77) describes Fine as holding the view that a 'property *F* is *essential* to an object *x* iff *x* could not have been the object it is without possessing the property *F*'. But what notion of possibility is being appealed to in this characterization? It had better not be the one characterized in terms of necessity in the standard way,[29] since otherwise we are back to our initial problem. For it is true that on this reading of the modal operator the singleton of Socrates could not have been the object that it is without having Socrates as a member. But, on the very same reading, it also seems true that Socrates could not be the object that it is without being an element of his singleton. The modality in question, then, has to be of a different sort, which makes it doubtful whether there is any substantial difference between the proposed analysis and taking the relation of dependence as primitive and unanalysed.

This is even clearer if we look at Fine's actual characterization of the notion of an essential property. For Fine actually refrains from giving the characterization in terms of possibility that Linnebo is attributing to him. Rather (1995: 273), he says that *x* is essentially *F* if and only if it is true in virtue of the identity of *x* that it is *F*. To this, however, he adds that

[a]lthough the form of words 'it is true in virtue of the identity of *x*' might appear to suggest an analysis of the operator into the notions of the identity of an object and of a proposition being true in virtue of the identity of an object, I do not wish to suggest such an analysis. The notation should be taken to indicate an unanalyzed relation between an object and a proposition. (Fine 1995: 273)

Thus, the notion of dependence is analysed in terms of the notion of an essential property of an object, which, in turn, is analyzed in terms of a primitive notion

---

[29] That is, by defining 'it is possible that *p*' as 'it is not necessary that not *p*'.

of being true in virtue of the identity of an object. But it is far from clear that we understand the unanalyzed relation 'it is true in virtue of the identity of x' any better than we understand the relation of metaphysical dependence itself. And hence it is also unclear that much is gained by taking the former, rather than the latter, as a primitive.

Similar considerations apply to Lowe's analysis. According to him, something is essential to x, if and only if, roughly, it is part of what x is (see Lowe 2005). It is hard to see to what extent we are told much more than that an object depends on another object just in case the identity of the former is determined by the latter.

The point is even starker if we look at Lowe's analysis from a slightly different perspective. Setting aside its recourse to essences, the account draws on the idea that an object depends for its identity on another object if and only if any individuation of the former must proceed via the latter (as Linnebo himself notices). But what is meant by 'individuation' here? Lowe characterizes it as

a certain kind of metaphysical determination relation between entities. In this sense, an entity – or, more specifically for present purposes, an object – is individuated by one or more other entities, its individuator or individuators. An object's individuators, in this metaphysical sense, are the entities that determine which object it is. (Lowe 2007: 521)

So, again, all we are told, effectively, is that an object depends for its identity on another object if and only if the latter determines which object the former is. And the problem is that it is hard to see what is gained by being told that 'y depends on x for its identity' is to be understood as meaning that x determines the identity of y.

It seems extremely doubtful, then, that these accounts can have the explanatory role they purport to have – or, at least, that Linnebo is attributing to them. Rather, all the accounts under consideration seem to rely, at bottom, on some primitive notion of metaphysical dependence, somehow arising out of the fact that a collection is constituted from or determined by its members, or has the members it has in virtue of its identity.

The main challenge for the dependency account, then, lies in making it plausible that we have replaced our metaphorical talk of set formation with something which can have a genuine explanatory role. The issue is particularly pressing because, according to the dependency theorist, the claim that the hierarchy covers all sets follows from the existence of the said relation of metaphysical dependence between a set and its members. But for this to be the case the relation of metaphysical dependence needs to have a certain number of structural properties.

For a start, it needs to be *irreflexive*, so that no set can depend upon itself. This would ensure that no set can feature among its own members, which, as we have seen, explains why there can be no universal set and no Russell set. But why should the dependence relation *be* irreflexive? If the substantial approach is to sanction an iterative set theory, rather than a set theory which admits of self-membered sets,[30] the dependency theorist has to provide an answer to this question, especially in the light of the fact that some of the proponents of the notion of metaphysical dependence accept that an entity can depend upon itself. For instance, Lowe argues that *everything* depends upon itself, since, he claims, for any entity x, there is a function, namely the identity function, such that it is essential to x that x is the entity identical with x.[31]

Moreover, for the hierarchy to cover all sets, the relation of metaphysical dependence has to be *antisymmetric*, so that although a set always depend upon its members, the members never depend upon the set. But again, the question is *why* we should think this to be the case, especially given that outside the metaphysical domain it is rather standard to claim that two things depend upon each other.

Against this, one might suggest that we have a good grasp of the notion of metaphysical dependence, and that this grasp is such as to make it *obvious* that metaphysical dependence is irreflexive and antisymmetric. But this flies in the face of the fact that, as we have seen, there are philosophers who have defended the view that, for instance, entities can metaphysically depend upon themselves, and that their views are not *obviously* wrong.

Ross Cameron (2008: 3), whilst agreeing with our diagnosis that metaphysical dependence cannot be analyzed in an informative way, claims that 'we have a good grasp of it; it is the relation that any impure set bears to the individuals in its transitive closure'. Now it might be granted that we have a good grasp of the relation that there exists between a set and its members if this is just the membership relation. And it might be granted that, in this case, if we take the membership relation to have certain structural properties, these structural properties will also be properties of the dependence relation. But now the problem is that we are using the notion of metaphysical dependence to motivate certain structural properties of the membership relation, and then using putative structural properties of the membership relation to motivate the structural properties of the notion of metaphysical dependence. It seems that we are going in a circle, and a very tight one indeed.[32]

---

30 In Chapters 6 and 7, we shall encounter set theories which, albeit consistent if ZF is, allow for sets of this kind.

31 Of course, as Lowe himself is eager to stress, endorsing the claim that everything depends upon itself does not commit one to the claim that everything depends *solely* upon itself.

32 To be sure, in the quoted passage Cameron is not dealing with the question whether the membership relation has certain structural properties. The point is just that, *if* someone wants to use the notion of metaphysical dependence to motivate certain structural features of the membership relation, they cannot rely on our

Maybe, though, the idea is that we should just *assume* that the dependence relation is irreflexive and antisymmetric, as well as transitive (see, e.g., Cameron himself 2008: 3 and Schaffer 2009: 364 and 376). This, however, makes the recourse to the notion of metaphysical dependence *redundant*, at least with regards to the attempt to motivate some of the structural features that the membership relation must have if the hierarchy is to cover all sets. For these structural features are motivated on the basis of the fact that sets depend upon their members, where it is just assumed that the dependence relation in question has certain structural features. But then, we might as well have assumed that the membership relation *itself* has these structural features, without any *detour* through the notion of metaphysical dependence: metaphysical dependence has become an idle wheel in our philosophy of set theory.

But if the dependency account is to vindicate the claim that the hierarchy covers all sets, it is not enough for the relation of metaphysical dependence to be irreflexive, antisymmetric and transitive: it also needs to be *well-founded*. This means that the dependency theorist needs an argument to the effect that all chains of metaphysical dependence terminate. And it is far from clear that there is any good argument to that effect.

It would take us too far afield to enter this debate here, but let me mention just one point in this connection. Cameron (2008) discusses some of the standard arguments against infinite descending chains of metaphysical dependence, and argues (persuasively, to my mind) that they are unconvincing. He then goes on to suggest that we should believe that every chain of metaphysical dependence terminates because 'it would be better to be able to give a common metaphysical explanation for every dependent entity' (2008: 12). Thus, he claims, we should accept the intuition that there is a fundamental level because, in so doing, we can 'give a better explanation of the phenomena to be explained' (2008: 13). For a start, this assumes that if x depends on y, then y explains x, which is far from obvious, albeit perhaps customary in the literature on metaphysical explanation. But in any case, it is unclear why being able to give a common explanation for every *dependent* entity (and not, crucially, entities belonging to the fundamental level – recall that Cameron assumes that the dependency relation is irreflexive) is better than being able, each time we are presented with any entity whatsoever, to provide an explanation for it.

So, to sum up, the account based on a primitive notion of metaphysical dependence seems to be the only real candidate for an account which attempts to argue

---

understanding of the membership relation to motivate the structural features of the relation of metaphysical dependence.

directly for the claim that all sets are those in the hierarchy. The other accounts either fail to sanction the axioms of set theory they purport to sanction or reduce to mere heuristics for those axioms. On the other hand, it is not clear to what extent we have a good grasp of the notion of metaphysical dependence, and, even if we do, there are major problems with showing that the claim that the iterative conception is correct about sets and levels follows from the account based on it.

2.5 Structuralism and Dependence

For these reasons, one might wonder whether other options are available to defend the view that sets are the objects that occur at some level of the cumulative hierarchy. A very different reason is offered by *mathematical structuralism.*

Very roughly, mathematical structuralism is the view that pure mathematics is the study of abstract structures. Two versions of structuralism can be distinguished. *Non-eliminative* structuralism takes talk of structures at face-value, and posits mathematical structures as genuine entities in their own right. It holds that mathematical objects are positions in these structures. *Eliminative* structuralism combines the insight that mathematical theories describe structures with an attempt to avoid commitment to an ontology of structures. It denies that the nature of mathematical objects is exhausted by their being positions in structures.[33]

One important issue for the non-eliminative structuralist is how her view differs from traditional mathematical platonism: after all, platonists too regard mathematics as dealing with abstract objects, and they will happily grant that mathematical objects participate in relations with other objects belonging to the same mathematical domain, so as to give rise to systems having a certain structure. To give an example, the arithmetical platonist believes that natural numbers exist, and that they form a system whose structure is that of the natural numbers. To be sure, it is the non-eliminative structuralist's contention that what mathematics is primarily concerned with is structures, not individual objects. But this makes it sound as though the difference between non-eliminative structuralism and platonism is only one of emphasis. Note that the issue does not obviously arise for *eliminative* versions of structuralism, which deny that the nature of mathematical objects is exhausted by their being positions in structures. For this reason, we shall restrict attention to non-eliminative versions of structuralism, and reserve the term 'structuralism' for those.

---

[33] See Linnebo 2008: 60–61 for more on the distinction, and for a careful description of different brands of non-eliminative structuralism.

The structuralist's usual response to this sort of worry is to remind us that in her view mathematical objects are *merely* positions in structures. However, this observation allows us to make some progress only insofar as it is conjoined with some account of how positions in structures differ from mathematical objects as conceived by traditional platonism. To this end, structuralists tend to make a number of additional claims about the nature of mathematical objects which, they contend, follow from their being positions in structures and which the platonist would reject.

Of particular importance among these claims is what Linnebo 2008 calls the *dependence claim*, which states that each mathematical object, *qua* position in a structure, depends on every other object in the structure and on the structure itself. The claim is endorsed by most structuralists; Shapiro, for instance, writes:[34]

The number 2 is no more and no less than the second position in the natural number structure; and 6 is the sixth position. Neither of them has any independence from the structure in which they are positions, and as positions in this structure, neither number is independent of the other. (Shapiro 2000: 258)

More recently, Shapiro (2008: 18–19) has emphasized that the dependence claim is the mathematical structuralist's main metaphysical thesis. And in the same vein, Linnebo (2008: 63–72) has argued that it is essential to the most plausible characterization of the structuralist's project.

What kind of dependence is involved in the dependence claim? Shapiro (2008: 18–24) has made it clear that the dependence in question is metaphysical and, notwithstanding some of his earlier pronouncements on the issue (see Shapiro 2006b), is not to be understood in terms of possible existence. So the structuralist too seems to face the issue of whether some notion of metaphysical dependence can be coherently made out. Note, incidentally, that the structuralist's commitment to the dependence claim implies that the dependence relation in question cannot be antisymmetric, at least as long as she holds on to the half of the dependence claim stating that every object in a structure depends on every other object in the structure. Perhaps this should cast further doubts on the idea that we can assume the relation of metaphysical dependence to have certain structural properties without further ado.

But the major issue arising in connection with the dependence claim is that, as has been pointed out by Linnebo (2008: esp. 72–74) and Parsons (2008: ch. 4), it gives rise, at least *prima facie*, to a tension with the dependency account of the iterative conception. For if this account is correct, a set depends on its members,

---

[34] See Linnebo 2008: 66–68 for evidence that Resnik too is committed to the dependence claim, albeit less explicitly.

but not vice versa, and, moreover, a set, far from depending on the whole of the cumulative hierarchy, *only* depends on its members. This would seem to show that we have a counterexample to the structuralist's thesis, at least if we hold on to the iterative conception.

Linnebo (2008: 77) concludes that 'the structuralists are wrong about (iterative) sets and that the scope of the dependence claim will have to be restricted accordingly': although in his view correct for structures like groups and rings, structuralism is untenable for the case of iterative set theory. This conclusion, however, relies on the assumption that the iterative conception is to be understood in terms of the dependency account. Mathematical structuralists, therefore, might well welcome an approach which offered a defence of the claim that the hierarchy covers all sets without requiring the existence of an antisymmetric relation of metaphysical dependence between a set and its members. Such an approach would enable them to hold on to the iterative conception whilst retaining the claim that each object in a structure depends on every other object in the structure and on the structure itself.

2.6 The Minimalist Account

The alternative approach in question is what I shall call the *minimalist* approach.[35] The minimalist approach consists of two main components: a view of how the iterative conception is to be understood, which one may call the *minimalist account* of the iterative conception; and a view of why the iterative conception so understood is to be regarded as correct, which, for reasons which shall later become clear, I will refer to as *inference to the best conception*. I shall present and discuss these two components in turn.

According to the minimalist account, the content of the iterative conception is exhausted by saying that it is the conception of set according to which sets are the objects that occur at one level or another of the cumulative hierarchy. The analogy here is with the case of the natural numbers: it is plausible to regard our conception of the natural numbers as exhausted by saying that numbers are the objects that occur at one point or another in the natural number series. Thus, just as on this view our conception of the natural number structure can be conveyed by saying that the natural numbers are 0, and its successor, and *its* successor, and *ITS* successor, and so on, on the minimalist account the iterative conception of the universe of (pure) sets can be conveyed by saying that the sets are the empty set and the set containing

---

[35] The approach might be anticipated by some brief remarks in Boolos 1989: 90–91 and Boolos 2000: 126–127.

the empty set, sets of those, sets of *those*, sets of *THOSE*, and so on, and then there is an infinite set, and sets of its elements, sets of those, sets of *those*, sets of *THOSE*, and so on, as far as possible; similarly for the case when individuals are admitted.

To pursue the analogy with the case of arithmetic a bit further, just as forming a conception of the natural number series requires a grasp of the successor operation, so forming a conception of the set-theoretic universe requires a grasp of the *set of* operation. The iterative conception, on the minimalist account, is then a conception of sets as sets *of* something: sets are what can be obtained from the empty set, or from the individuals, by iterated applications of the *set of* operation. This gives rise to the cumulative hierarchy structure since, as Gödel himself noticed (*1951: 306, fn. 5), the *set of* operation, at the formal level, is 'substantially the same' as the powerset operation.[36]

Recall that for the constructivist sets really are formed in stages. For the dependency theorist, on the other hand, talk of set formation is at best a metaphor to be replaced by a relation of metaphysical dependence holding between a set and its members. But for the minimalist, talk of set formation arises out of a narrative convention whereby 'in general and *ceteris paribus*, a description of objects that are arranged in some salient manner should mention those objects in an order corresponding to the arrangement' (Boolos 1989: 90). So talk of set formation is neither to be taken literally nor is it a metaphor for some notion of metaphysical dependence. Rather, it is due to this kind of narrative convention, and can be removed at the expense of simplicity of exposition. The point is made by Boolos with his characteristic verve:

In any case, for the purpose of explaining the conception, the metaphor is thoroughly unnecessary, for we can say instead: there are the null set and the set containing just the null set, sets of all those, sets of all *those*, sets of all *Those*, ... There are also sets of all *THOSE*. Let us now refer to these sets as 'those'. Then there are sets of those, sets of *those*, ... Notice that the dots '...' of ellipsis, like 'etc.,' are a demonstrative; both mean: *and so forth*, i.e. in *this* manner forth. (Boolos 1989: 91)

This, the minimalist claims, makes it clear that in order to grasp the cumulative hierarchy picture of the set-theoretic universe we need not think of sets as metaphysically dependent upon their members, just as in order to form a conception of the natural number series we need not think of natural numbers as metaphysically dependent upon their predecessors. What the minimalist claims, in addition to this,

---

[36] Elsewhere, Gödel (1947: 180, fn, 13) also points out that albeit the expression 'set of *x*'s' cannot be defined away, it can be paraphrased by other expressions such as 'combination of any number of *x*'s', of which, he says, it is reasonable to claim that we have an independent grasp. Note, incidentally, that Gödel's explanation of the *set of* operation makes the iterative conception a combinatorial conception.

is that *all there is* to the iterative conception is that sets are the objects which occur at some level of the cumulative hierarchy. This means that, on the minimalist account, it is not part of the iterative conception that sets are metaphysically dependent upon their members. Hence, it also follows that by endorsing this account the structuralist becomes free, if she wishes, to say that every set in the cumulative hierarchy structure metaphysically depends on every other set in the structure and on the cumulative hierarchy itself.

One might worry that the minimalist account is not what we *mean* by 'iterative conception': the iterative conception, the worry goes, is *precisely* the conception of set according to which sets are somehow constituted by their members, and so upholding the minimalist account amounts to changing the subject. There are two points to be made in response to this worry.

The first point is that the minimalist account, albeit neglected in the current literature, does have the credentials to be a faithful account of what, historically, has often been meant by 'iterative conception'. To give but one example, the minimalist account is arguably the one which Gödel has in mind in the paper that is considered by some the paper in which the iterative conception is described for the first time in print (see, e.g., Potter 2004: 36). He writes:

This concept of set, however, *according to which a set is anything obtainable from the integers (or some other well-defined objects) by iterated application of the operation ("set of",)* and not something obtained by dividing the totality of all existing things into two categories, has never led to any antinomy whatsoever; that is, the perfectly ("naïve") and uncritical working with this concept of set has so far proved completely self-consistent. (Gödel 1947: 180, emphasis mine)

Thus, Gödel himself, in line with the minimalist account, describes the iterative conception as the conception of set according to which sets are what can be obtained from the individuals by iterated applications of the *set of* operation.

The second point is that, in any case, the minimalist account is not a view of what we *mean* by 'iterative conception'; rather, it is a view of what we *ought* to mean by it. The minimalist is pointing out that there is an understanding of the iterative conception which can do all the work that the iterative conception was invoked to do without postulating the existence of a dependence relation between a set and its members. In particular, recall the two aspects of the conception that make it so appealing: first, it offers a picture of the set-theoretic universe which fits well with a highly plausible diagnosis of the set-theoretic paradoxes and explains the non-existence of, e.g., the universal set and the Russell set; second, it sanctions (at least) most of the axioms of standard set theory. These two aspects of the conception continue to hold even on the minimalist approach, since they only depend on the cumulative hierarchy picture of the set-theoretic universe, which is retained on the minimalist account. The appeal of the dependency account was that it promised to explain *why* sets are arranged in a cumulative hierarchy divided into levels. But, as we have seen, the account seems unlikely to be able to fulfill such a promise. This, the minimalist claims, shows that the relation of metaphysical dependence is really not doing any work, and should therefore drop out of our account of the conception.

All of this brings out an analogy between the minimalist account of the iterative conception and the minimalist account of *truth*. (The analogy is non-coincidental, and deliberately evoked by the similarity of the two names.) Take Paul Horwich's (1998) account, for instance. He holds that a theory of truth should comprise no more than the (non-paradoxical) instances of the Equivalence Schema

$$\text{The proposition } \textit{that } p \text{ is true iff } p, \tag{E}$$

since all facts about truth can be explained by appeal to it, and that it is unlikely that they can be explained in terms of anything more fundamental.[37] Similarly, the minimalist account of the iterative conception holds that the content of the conception is exhausted by saying that sets are the objects that occur at one level or another of the cumulative hierarchy, since this suffices to explain all facts about sets that the conception was invoked to explain – that is, why the problematic collections involved in the paradoxes do not exist and how, nonetheless, (most of) the axioms of standard set theory can be true. And, the minimalist claims, it is unlikely that these facts can be explained in terms of something more fundamental such as the relation of metaphysical dependence.

2.7 Inference to the Best Conception

So far the minimalist has not explicitly addressed the question we started with: what reasons do we have for thinking that the iterative conception is correct? The minimalist does not try to argue that all sets are those in the hierarchy by attempting to show how this follows from some very general considerations about the nature of sets. In fact, as we have been stressing throughout, one possible reason for adopting a minimalist approach is precisely that such a strategy is marred with difficulties, and the minimalist account explicitly refrains from making any substantial demand

---

[37] To be precise, this is Horwich's view as expressed in Horwich 1990, the first edition of Horwich 1998. In response to criticisms, Horwich has slightly modified his view about what the minimal theory of truth should comprise. However, the point is irrelevant for our present concerns, since his reasons for taking something to be part of the minimal theory remain the same. For details, the reader can consult the postscript to the second edition of Horwich's book.

on the metaphysics of sets. As anticipated above, however, there is a second component to the minimalist approach, which represents the minimalist's view of why the iterative conception, as she understands it, is to be regarded as correct.

According to this view, we should regard the iterative conception as correct because it is the most satisfactory conception of set that we have – that is, because it is better than its rivals with regards to certain virtues that a conception of set may have. This view arises quite naturally out of the minimalist's emphasis that the appeal of the iterative conception is due to the fact that it provides a convincing explanation of the paradoxes whilst showing how (most of) the axioms of standard set theory can be true. These two virtues, the minimalist thinks, are themselves part of the reason why we should regard the iterative conception as correct. To be sure, the minimalist does not want to say that the conception is correct *only* because it has these two virtues. For a start, there might be *other* conceptions of set that are equally capable of providing a convincing explanation of the paradoxes whilst sanctioning theories strong enough that it is possible to develop a substantial amount of mathematics within them. What the minimalist wants to say, rather, is that these and perhaps other virtues of the iterative conception make it the most satisfactory conception of set that we have, and for this reason the conception should be regarded as correct.

Thus, it is a consequence of this aspect of the minimalist approach that the arguments for the iterative conception are partly going to be indirect: the iterative conception is to be defended by showing that the other conceptions of set available are unsatisfactory, or at least less satisfactory than the iterative conception. To use a slogan, the argument for the iterative conception is an *inference to the best conception*.[38] Hence, a full defence of the iterative conception will involve looking at the viability of conceptions of set other than the iterative one, together with, presumably, some discussion of the virtues a conception of set may have, i.e. of the criteria which may be used to show a conception to be more satisfactory than its rivals. These tasks will occupy us in the remainder of the book. For the moment, let me say something on behalf of the two aspects which, I have claimed, make the iterative conception initially so appealing and will play an important role in its defence. Why should these aspects be taken as criteria for deeming a conception to be satisfactory?

Being able to explain the paradoxes can be taken as such a criterion because consistency in a theory is a virtue. This latter claim is fairly uncontroversial, and

---

[38] The analogy, of course, is with inference to the best *explanation*, the *locus classicus* for which is Lipton 2003.

indeed is usually endorsed even by philosophers and logicians who attempt to rehabilitate the naïve conception by developing corresponding set theories which, albeit inconsistent, are non-trivial because the underlying logic is non-classical (see Chapter 4). Where these people disagree from the defender of the iterative conception is in claiming that this virtue, in the case of set theory, is trumped by other virtues. But once it is recognized that consistency in a theory is a virtue, the need arises for some *reason* for thinking that our theory has this virtue. And in the case of set theory, we want this reason to be different from the fact that the theory can be proved consistent in a stronger background set theory in the metalanguage, since this only shifts the problem one level up. The iterative conception provides us with such a reason, since it is easy to see that, for instance, the universal set and the Russell set occur at no level of the cumulative hierarchy.

What about being able to sanction a theory in which it is possible to carry out a substantial amount of mathematics? This criterion too should not be very contentious once it is considered what set theory has often been *for*. Take, for instance, two of the goals which have been historically important in the development of set theory. The first is the goal of giving a foundation for classical mathematics, in the sense of providing

a dependable and perspicuous mathematical theory that is ample enough to include (surrogates for) all the objects of classical mathematics and strong enough to imply all the classical theorems about them. (Maddy 2007: 354)

This goal is accomplished, at least in part, by showing how arithmetic and analysis can be embedded in set theory. Thus, whether or not the foundation provided by set theory is going to be a foundation for *classical* mathematics, a certain amount of set-theoretic machinery is likely to be required to accomplish this goal.[39] The second goal set theory has often had is that of providing a tool for understanding the infinite, and this goal too seems to require that our set theory have certain resources such as those needed to develop interesting and fruitful theories of cardinals and ordinals. Thus, both goals depend on the strength of our set theory: if set theory is not capable of accomplishing these aims, it seems deprived of a large part of its interest and use. This offers a reason for thinking that whether a conception sanctions a set theory which is strong enough to allow for the development of certain set-theoretic constructions should count against or in favour of that conception.

---

[39] See Shapiro 2004 for different senses in which set theory has been thought to be a foundation for mathematics, e.g. metaphysical or epistemological. In these senses too, set theory needs to retain some mathematical strength if it is to serve as a foundation.

But if the evaluation of a conception depends on what kind of set-theoretic constructions one can carry out in the set theory it sanctions, the development of theories becomes an integral part of the assessment of conceptions. This gives rise to a position concerning the justification of the axioms of set theory which is worth contextualizing.

Following Gödel (1947; 1964), it is customary to distinguish between two kinds of justification of the axioms of set theory, *intrinsic* and *extrinsic*. Roughly, intrinsic justifications are based on conceptions of set, whereas extrinsic justifications are based on the success of those axioms. Whether an axiom is justified *on* a certain conception is going to be based on the content of the conception and is therefore going to be an intrinsic justification. But whether the axiom is justified *by* the conception is also going to depend on whether the conception is correct, and that, in turn, also depends on the success of the conception for the development of mathematics. Thus, there is an extrinsic component to the justifications offered in this book. To stress, however: the success of a conception for the development of mathematics is only *one* among the desiderata that one might have on conceptions of set.[40]

Some methodological considerations are in order. In this chapter we have been distinguishing between the question of which axioms the iterative conception sanctions and the question of whether the iterative conception is correct. This is in line with what we suggested in the previous chapter (see Section 1.7), and enables us to separate two tasks which are in principle different, that is, the task of showing which axioms are justified on a certain conception, and the task of showing whether a certain conception is correct or not, although, as observed, *which* axioms follow from a certain conception might itself be relevant to determining whether the conception is satisfactory or not.

Albeit intuitive and plausible, the distinction we have been drawing is sometimes absent in treatments of the iterative conception. For instance, Paseau (2007: 37–41) discusses Boolos on the justification of the axioms of set theory (see Boolos 1971; 1989). To this end, he lays down four criteria of justification which he traces back to Boolos's text. Suppose we want to ascertain whether and to what extent a certain conception C or a certain principle P constitutes an adequate justification for a certain axiom A. Then, according to Paseau's reading of Boolos, we need to ascertain, beyond mere logical consequence, how many of the following criteria are satisfied:

---

[40] Remarkably, Gödel himself seems to have held that there is an extrinsic element in the assessment of the tenability of a conception. See Parsons 2014.

(1) The background conception C must be natural and consistent.

(2) The background conception must be an actual conception.

(3) To decide whether P is part of C must be an easier decision than whether to accept A.

(4) The obviousness of P must not be independent from that of the rest of C.

However, if one differentiates between an axiom being justified on a conception and the conception being correct, it becomes philosophically important to keep criteria (1) and (2) distinct from criteria (3) and (4): criteria (1) and (2) could be useful to assess whether C is correct; criteria (3) and (4) could be used to assess whether a certain theory or a certain axiom is justified on C. Moreover, there is strong textual evidence that this is the approach that Boolos himself is urging.

In his 1989 paper, Boolos says:

The iterative conception is the only natural and (apparently) consistent conception of set we have, and it implies $[Z_i]$; *that* is the justification (if that is the right word) it provides for $[Z_i]$. (Boolos 1989: 90)

In this passage Boolos seems precisely to be endorsing the view I am suggesting: for some axioms to be justified, it is not sufficient that they follow from a certain conception (or some formalization thereof); the conception must also be correct – where Boolos's criteria for correctness are naturalness and consistency. Later in the book, we will cast doubts on Boolos's claim that the iterative conception is the only 'natural and (apparently) consistent conception of set that we have' (see Section 7.3). Here, what Boolos says is clearly at odds with Paseau's interpretation, since, on Paseau's view, an axiom can be justified even if a certain conception is not natural (provided that some of the criteria (2)–(4) are satisfied). Paseau concludes that since the passage 'undermines the main points and most of the content of his two articles, the reasons for disregarding it are very strong' (2007: 39), and that 'Boolos's characterization of his project in the specific passage on p. 90 is mistaken' (2007: 40, fn. 14). But what are the reasons for disregarding it? In the footnote just mentioned, Paseau offers as evidence the fact that, in his paper, Boolos says that another conception we shall discuss in Chapter 5, the limitation of size conception, 'accounts for' and is 'behind' some of the ZFC axioms, which are also said to 'follow from' this conception. But this is perfectly compatible with the interpretation I am putting forward: a set of axioms can follow from, be sanctioned by or be justified on a conception which is not correct. But a set of axioms can only be justified *tout court* by a correct conception, and Boolos obviously thinks that a conception which is not natural cannot be correct.

Similar considerations to those offered against Paseau's reading could be used to defuse an objection raised by Oliver (2000: 866) against Boolos's claim that some of the ZFC axioms follow from the iterative conception, whereas some of the others follow from the limitation of size conception. Oliver's objection is that this commits Boolos to theories of numbers at odds with each other. The way out of the objection should now be clear: Boolos thinks that some of the ZFC axioms, namely Extensionality, Replacement and Choice follow from the limitation of size conception, but not from the iterative conception. However, he does not think that the limitation of size conception is correct. Hence, he is not committed to what follows from the limitation of size conception.

2.8 Conclusion

It is often thought among philosophers that the idea that there exists a relation of priority or dependence between sets and their members is part and parcel of the iterative conception. The main contention of this chapter has been that this is mistaken: it is possible to understand the iterative conception as making no demands on the metaphysics of sets but only on the structure of the system to which they give rise. To be sure, there might be other reasons for regarding sets as dependent upon their members, but I hope at least to have shown that these reasons do not come from within set theory.[41] For there is a minimalist account of the iterative conception which makes do without notions such as priority or metaphysical dependence. This account, moreover, provides the structuralist with the resources to hold on to the iterative conception whilst retaining the claim that each object in a structure depends on every other object in the structure and on the structure itself. Of course, as I have pointed out, the minimalist approach has the consequence that a defence of the idea that the hierarchy covers all sets requires one to show that the rivals of the iterative conception are less satisfactory than the iterative conception itself. This is the task that we shall undertake in the remainder of the book.

---

[41] Some arguments to this effect are offered in Fine 1994. For the record, I do not find these arguments very convincing, but, for reasons of space, I cannot discuss them here.

3

Challenges to the Iterative Conception

> *O, thou hast damnable iteration,*
> *and art indeed able to corrupt a saint.*

> *William Shakespeare,* Henry IV

In Chapter 2, I introduced the iterative conception of set and explained some important differences between various accounts thereof. These differences notwithstanding, all accounts share the idea that the set-theoretic universe is a cumulative hierarchy divided into levels. Now the development of mathematics and mathematical logic during the last century has made it clear that the cumulative hierarchy is a perfectly respectable mathematical structure, which we now know a great deal about. Few people are willing to dispute this. What is usually disputed is the claim that the cumulative hierarchy comprises all sets.

Some people limit themselves to saying that there is no good *argument* for this claim: there is just no reason to suppose that all sets are those in the hierarchy. I will discuss this kind of position in Chapter 7, but recall that, on the minimalist approach, the claim that the hierarchy covers all sets is to be defended by showing that the iterative conception of set is the most satisfactory conception of set that we have.

In this chapter, on the other hand, we will examine a number of objections to the very idea that sets are the objects which appear at some level of the cumulative hierarchy. More specifically, we will discuss arguments to the effect that if we take all sets to be those in the hierarchy, then we have to face a series of unacceptable consequences. Unsurprisingly, they are advanced by people who advocate a conception of set other than the iterative one, usually the naïve conception (see especially Weir 1998a; Priest 2006b). Thus, it is claimed, if there are strong arguments against the idea that all sets are those in the hierarchy, we should reconsider the viability of the naïve conception. I will discuss the prospects for this project in the next chapter. My main aim in this chapter is to review arguments against the idea that all sets are those in the hierarchy and show that they fail to undermine the iterative conception. They can be summarized as follows:

1. If we take all sets to be those in the hierarchy, we cannot explain the appeal of the naïve conception of set.
2. If we take all sets to be those in the hierarchy, we cannot give a non-circular account of the cumulative hierarchy.
3. If we take all sets to be those in the hierarchy, we cannot give a semantics for set theory.[1]

I am sure there must be other possible objections, but these are the ones that are usually advanced by critics of the iterative conception.[2] I shall take them in order. The chapter ends by discussing the status of Replacement on the iterative conception.

3.1 The Missing Explanation Objection

The first objection, which I shall call the *missing explanation* objection, is that if we take all sets to be those in the hierarchy, we cannot explain the appeal of the naïve conception of set, as embodied in the Axioms of Comprehension and Extensionality. And, the objection goes, any respectable conception of set should be capable of explaining why people tend to find these principles so compelling. People who advance this objection usually contrast the situation with that of the naïve conception of set. This conception, they argue, *can* explain why we find the Axioms of Comprehension and Extensionality so compelling – they are, after all, true. Priest sums up the objection as follows:

---

[1] Priest (2006b) formulates the objections by saying that if all sets were those in the hierarchy, then a number of unacceptable consequences would follow. To avoid entering the debate on the truth-value of counterpossibles, I have reformulated the statement of the objections.

[2] To be sure, Priest (2006b: 32–35) also advances another objection to the iterative conception, to the effect that if we take all sets to be those in the hierarchy, then we cannot take set theory to have a foundational role because of its alleged inadequacy in dealing with category theory and in particular categories such as the category of all sets or the category of all categories. For reasons of space, I cannot address this objection here, but let me say two things about it. First, although some of the strategies for dealing with the problem are admittedly unsatisfactory, the prospects for overcoming it by interpreting category theorists as making statements having a schematic character are not as dim as Priest makes them out to be (see Awodey 2004; Feferman 2004). Second, suppose that we do stick with the idea that talk about, say, the category of all sets is to be reconstructed as talk about the set of all sets. Then, the problem of providing a foundation for category theory is not solved simply by switching to set theories which allow for a universal set (see Chapter 6), as is witnessed by the difficulties encountered by attempts to reconstruct category theory in some of these set theories (see, e.g., McLarty 1992; Thomas 2018). Priest thinks that his theory based on the naïve conception has the resources to reconstruct category theory within it, but he offers no evidence for this, and, in fact, the limitations of his theory which we shall point out in the next chapter make it very doubtful that it does. For these reasons, it is far from clear that set theories based on the naïve conception or other conceptions of set are better off than those based on the iterative conception when it comes to dealing with category theory.

If the only instances of [the Comprehension Schema] that are true were those that held in the cumulative hierarchy, our naïve supposition that all conditions do define sets would appear inexplicable. From this perspective, it does not even seem plausible. Thus, the naïve theory, which does explain why we believe this (it is, after all, true) exceeds the theory based on the cumulative hierarchy in explanatory power. (Priest 2006b: 31)

Now, contrary to what this passage might suggest, the mere truth of the naïve theory does not suffice to explain why we naïvely tend to believe that the theory is true. After all, the truth of a proposition p is not generally considered sufficient to explain the fact that we believe that p, and is usually coupled with some account of how we arrived at our belief. It could be granted, however, that taking the supposition that all conditions define sets to be true puts the naïve theory in a position to provide the required explanation. And if that is right, the missing explanation objection might retain its force: by taking the supposition that every condition determines a set to be false, it could be argued, the theory based on the hierarchy is incapable of giving an explanation which can be offered by the naïve theory.

Again, however, this cannot be all there is to the objection: it is not generally true that we cannot explain why we believe that p if we take p to be false (of course). Presumably, this is why the objection is glossed over by saying that, if all sets were those in the hierarchy, it would not even seem *plausible* that every condition determines a set. But why is that? Consider what is probably the standard view on the matter – or at least the standard view of those who claim that the hierarchy covers all sets. The view is that the naïve supposition that every condition determines a set is just one of our untutored false suppositions or beliefs. (This is not to say that we cannot make use of *tutored* common sense in our reasoning – the intuitive arguments for the iterative conception are just a case in point.) As it stands, however, the standard view does not quite deal with the objection, since it still leaves it unexplained why we make the particular untutored and false supposition that all conditions determine a set. Compare this with the case of false beliefs such as the belief that the Sun really rises: it is not enough to say that this is one of our false untutored beliefs; we also want to explain why people have tended to have that belief, and we can do this, for instance, by pointing out facts about perception and the orbits of the Sun and the Earth.

So the objection could be that there is nothing analogous to appeal to in this case – nothing which makes the supposition that every condition determines a set even just plausible (to us) whilst being compatible with the claim that all sets are those in the hierarchy. Against this, I want to suggest that there are at least two possible explanations of why people tend to find the naïve supposition that all instances of the Comprehension Schema are true initially so appealing.[3] Both explanations, at least in some versions, share the idea that what we are doing if and when we make the naïve supposition that all conditions determine a set is to conflate two different concepts/conceptions, one of them being the concept of set (as sharpened by the iterative conception of set), the other one being a concept/conception whose truth is compatible with the iterative conception.[4] I shall present them in turn.

Sets of Individuals and Sets of Sets

Consider one possible way of motivating the naïve conception. The idea is that, for any condition φ, we can collect together all and only those things which satisfy φ. As Boolos (1971: 14) puts it, 'How could there *not* be a collection, or set, of just those things to which any given predicate applied?' This is often taken to be the informal reasoning that leads us to believe that the Naïve Comprehension Schema is true.

Notice, however, that talk of 'things' seems to leave it indeterminate whether we are talking only of individuals or also of sets. This, however, is crucial because if we take 'things' to be only individuals and not also sets, we end up with a conception of set – call it the *naïve conception of set of individuals* – which seems to sanction a theory whose truth is compatible with the claim that the hierarchy covers all sets. Indeed, the theory which the naïve conception of set of individuals seems to sanction is just a notational variant of an easy consequence of any impure iterative set theory which takes the individuals to form a set.

For according to the naïve conception of set of individuals, to every condition there corresponds the set of the individuals satisfying that condition. This seems to sanction the following modification of the Naïve Comprehension Schema:

$$\exists y \forall x (x \in y \leftrightarrow (U(x) \wedge \varphi(x))), \tag{3.1}$$

---

[3] There might seem to be something unsatisfying about the fact that we are offering more than one possible explanation. After all, it might be said, isn't there bound to be a single explanation of our naïve supposition that all conditions determine a set? This, however, is due to the nature of the missing explanation objection, and, if anything, is a problem for the proponent of the objection. For any explanation that the defender of the iterative conception can offer to answer the missing explanation objection can only be an *attempted* explanation until it is confirmed by some experimental data. These, presumably, will have to confirm: first, that people really do tend to make the naïve supposition that all conditions determine a set; and second, that the proposed explanation is supported by the evidence.

[4] Both explanations have been inspired by the following remark by Kreisel (1967: 82): '*class* presented itself as a vague notion, or specifically a mixture of notions including (i) finite sets of individuals [...], or (ii) sets *of* something (as in mathematics, sets of numbers, sets of points), but also (iii) properties or *intensions*, where one has no a priori bound on the extension (which are very common in ordinary thought but not in mathematics)'.

where $U(x)$ abbreviates 'x is an individual' and the formula $\varphi(x)$ satisfies the restrictions indicated for (Comp). Given this modification of the Naïve Comprehension Schema, the reasoning that leads to Russell's Paradox in the case of the standard Comprehension Schema simply becomes a proof that the set $u = \{x : U(x) \wedge x \notin x\}$ is not an individual. For suppose that it is. Then, by the defining condition of membership in u, we have that if $u \notin u$ then $u \in u$. On the other hand, if $u \in u$ then $u \notin u$. Hence, we have that

$$U(u) \rightarrow (u \in u \leftrightarrow u \notin u),$$

from which we can conclude that u is not an individual.

But now note that the thought that to every condition there corresponds a set of all and only those individuals satisfying that condition is part of the iterative conception, at least if we assume that the individuals comprise a set. For if we do, the iterative conception tells us that level 0 of the cumulative hierarchy will be the set of all individuals. And then, at level 1, we can form the set of all and only those things occurring at level 0 and satisfying condition φ. From a formal point of view, this is reflected in the fact that what is essentially a notational variant of (3.1) is a simple consequence of most iterative set theories which countenance individuals. For, in most theories of this kind, we can prove that there is a set of all individuals. Thus, we have the following instance of the Axiom of Separation,

$$\exists y \forall x (x \in y \leftrightarrow x \in \{z : U(z)\} \wedge \varphi(x)). \tag{3.2}$$

And from this, we can prove the existence of any set of individuals for any condition φ. Again, it is a familiar point that Russell-type reasoning leads to the conclusion that the set $u = \{x : x \in \{z : U(z)\} \wedge x \notin x\}$ cannot belong to $\{z : U(z)\}$, on pain of contradiction, which is just another way of putting the idea that on the iterative conception indefinite extensibility is retained at the expense of universality (see Section 1.7).

The basis of the proposed explanation of why people tend to find compelling the supposition that all conditions determine a set, therefore, is that they tend to think of a set as the result of collecting together *things*, by which they usually mean *individuals*. This thought is quite innocent, and indeed sanctions an axiom schema which is just a notational variant of an instance of the Separation Axiom. What we need, therefore, is some reason why this should lead us to thinking that every condition determines a set. One possibility is that we simply overlook the fact that, in the Naïve Comprehension Schema, *sets themselves* are among the things that can be collected together to form a set. Another possibility is that the supposition arises because we mix together two conceptions of set, the naïve conception of set of individuals and the iterative conception. For the latter is precisely the notion of sets as sets *of* something, including sets themselves. Thus, in supposing that to every condition there corresponds a set of all the things satisfying that condition – no matter whether these things are individuals or sets – we are conflating ideas from two conceptions, one of which is the iterative conception, another of which can just be regarded as a consequence of it. In either case, we have the basis of an explanation of why we naïvely tend to suppose that every condition determines a set: we are misled by the existence of a true thought in the vicinity, that is, the thought that to every condition there corresponds a collection of the things that satisfy that condition, as long as by 'things' we mean *individuals*. We then either conflate this thought with the notion of sets as sets *of*, or simply overlook the difference between this thought and the Naïve Comprehension Schema.

Sets and Objectified Properties

According to the second explanation, similarly to one version of the first explanation, our supposition that every instance of the Comprehension Schema holds is the result of conflating two different concepts, the concept of set and the concept of an objectified property – an object systematically associated with a property (see Section 6.6 for more details). Sets, as characterized by the iterative conception, are combinatorial collections; objectified properties are logical collections. Recall that on a combinatorial conception, membership in a collection is not derivative upon having a certain property. On a logical conception, it is: collections are intimately connected with properties. However, assuming that to every property there corresponds a collection of all and only the things having that property is classically problematic because of Russell's Paradox. For in the presence of (Comp2), it is enough to consider the extension of the property *does not fall under the concept of which it is the extension* and ask whether its extension has the property to obtain a contradiction (see Section 1.7).

Now recall again the diagnosis of the paradoxes that we offered in Chapter 1. According to that diagnosis, the paradoxes arise because on the naïve conception of set we have both indefinite extensibility and universality. The solution offered by the iterative conception is to reject universality and endorse indefinite extensibility. The fact that we reject universality for sets, however, does not mean that we have to reject it for objectified properties. Indeed, many have felt that it would be problematic to ban properties like *self-identity*, since the predicate 'is identical to itself' is as intelligible as any, and similar considerations would seem to apply for the entities that are taken to be systematically associated with those properties. For instance, John Myhill reports that

Gödel said to [him] more than once, 'There never were any set-theoretic paradoxes, but the *property-theoretic* paradoxes are still unresolved'. (Myhill 1984: 129)

Thus, the other possible explanation for the fact that we make the naïve supposition that all conditions determine a set is that we are conflating two concepts, the concept of set and the concept of objectified property. The former satisfies indefinite extensibility but not universality. The latter might satisfy universality and need not satisfy indefinite extensibility. Sets are combinatorial collections arranged in a hierarchy. Objectified properties are logical collections and need not be hierarchically arranged.

In passages quoted by Wang (1974: 85–86), Gödel too argues that paradoxical results can be arrived at when two or more conceptions of the concept at hand are conflated. Wang mentions, for instance, the case of the intuitive concept of continuity. He says that our prior intuition – what we have called the concept of continuity – 'contains an ambiguity between smooth curves and continuous movements'. Results concerning, for instance, space-filling continuous curves appear paradoxical because two conceptions of continuity are being conflated in the intuitive concept. Once the two conceptions are clearly distinguished, and it is realized that a space-filling continuous curve is only continuous in the sense of being smooth, the existence of such a curve appears no longer paradoxical. Similarly, in the case at hand, we must distinguish between two conceptions of collection: a combinatorial one, exemplified by the iterative conception, and a logical one. Once the two conceptions are clearly distinguished and we consider them as conceptions of different concepts, that is, the concept of set and the concept of an objectified property, we can explain the appeal of the Naïve Comprehension Schema without having to accept its truth for sets.

A further question which then arises is whether *indefinite extensibility* should hold for objectified properties. Classically, this cannot be the case, unless we want our theory of objectified properties to be trivial. So we seem to have two options. The first would be to retain indefinite extensibility and develop the theory of objectified properties in a different logic. This option is developed, for instance, by Hartry Field (2008), who, incidentally, seems to share the flavour of our explanation of why we naïvely tend to suppose that all conditions determine a set. Having noticed that Russell originally formulated his paradox in connection with Frege's theory of extensions, he writes that

[e]xtensions were supposed to be like mathematical sets in obeying an axiom of extensionality: no two extensions were instantiated by exactly the same things. But they were to be like conceptual properties in that every predicate had one, and that they didn't fall into a hierarchy of ranks. (Field 2008: 3)

Of course, we are not suggesting here that Frege himself supposed that all conditions determine a set because, effectively, he was conflating two different concepts, the concept of a set and the concept of an objectified property. But I take it that the missing explanation objection concerns why *people in general* tend to find the Naïve Comprehension Schema quite intuitive. The answer to the question why Frege *himself* made the supposition that all conditions determine a set is probably going to invoke certain facts about Frege's own project and philosophical system.[5]

The idea that to every predicate there corresponds an objectified property means that objectified properties, besides satisfying universality, also satisfy indefinite extensibility, at the price of abandoning classical logic. The second option, which, to my knowledge, has never been explored, would be to retain classical logic and revise our theory of objectified properties so that universality is retained and indefinite extensibility rejected. As we will see in Chapter 6, one way to develop this second option gives rise to a theory of objectified properties looking like Quine's system NF.

Gödel himself, when discussing the intensional paradoxes – differing from the set-theoretic paradoxes in that they concern properties rather than sets – considered the possibility of rejecting the Comprehension Schema for objectified properties and thought that some of the options which one might find unattractive in the case of sets might be less so in the case of something like objectified properties, which he called 'concepts' (see Section 1.7, fn. 10). So, for instance, Wang reports him as saying that, by distinguishing concepts from sets

we acquire not only a fairly rich and understandable set theory but also a clear guidance for our search for axioms that deal with concepts generally. We can examine whether familiar axioms for sets have counterparts for concepts and also investigate whether earlier attempts (e.g. in terms of the lambda-calculus and of stratification, etc.), which deal with sets and concepts indiscriminately, may suggest axioms that are true of concepts generally. (Wang 1996: 277)

In any event, the defender of the iterative conception need not, for present purposes, settle for one option or another. For what matters in the present circumstances is that both options can form the basis for another possible explanation of why we tend to find appealing the naïve supposition that all conditions determine a set: in making that supposition, we are conflating the concept of set, which as sharpened by the iterative conception is best regarded as combinatorial in character, and that of objectified property, which is logical in character. This explanation, moreover, fits well with the diagnosis of the paradoxes of Chapter 1: conflating the concept of set and that of objectified property leads to paradox because the

---

[5] For an explanation along these lines of why Frege fell into the contradiction, see Sullivan 2007.

former satisfies indefinite extensibility and the latter satisfies universality and the conjunction of these two features is inconsistent (see Theorem 1.7.1). We will return to this extended diagnosis of the paradoxes in Chapter 6.

3.2 The Circularity Objection

Let us now turn to the second objection. This is to the effect that if we take all sets to be those in the hierarchy, we cannot provide a non-circular account of the hierarchy. The objection, which we may call the *circularity* objection, is advanced by Priest in the following paragraph:

[I]t is not even clear that the notion of the cumulative hierarchy makes sense without a prior and different notion of set. For it is clear that the construction is parasitic upon a prior notion of an ordinal. [. . .] Given a (naïve) notion of set, this can be used to specify a theory of ordinals, and hence define the cumulative hierarchy, but there seems no other adequate way of characterising the ordinals. Hence the cumulative hierarchy seems to presuppose a different notion of set. Could the notion of ordinal be provided by a theory such as ZF? Not without damaging circularity, since this is one of the very set theories whose rationale pegs it to the cumulative hierarchy. (Priest 2006b: 31)

The objection is based on the fact that it appears that in order to fully comprehend the cumulative hierarchy of sets $V = \bigcup_\alpha V_\alpha$, we need to have already a prior understanding of the notion of ordinal. Since ordinals are generally described as order types of well-ordered sets and often identified with certain sets themselves (see Section 1.A), it would appear that our ability to understand the ordinals relies on our antecedent familiarity with order types of well-ordered sets. But this familiarity is surely mediated by set theory. The problem, then, is that this dependence of our understanding of the hierarchy on the notion of an ordinal would seem to compromise the ability of the iterative conception to provide a justification for the axioms of set theory. What we need for this purpose, Priest is urging, is another, independent conception of set – the naïve conception. With this more basic and different conception on board, we would be able to grasp the cumulative hierarchy. But its justificatory role with respect to the axioms of set theory would be jeopardized.

The objection is familiar to philosophically minded set-theorists, who directed their attention to it from the early days of axiomatic set theory. For instance, a version of the objection is echoed in the following remark by Gödel, dating as far back as 1933:[6]

---

[6] More recently, the worry has also been voiced by Antonelli (1999: 150), who quotes approvingly the following remark by Fraenkel et al. (1973: 89): 'The weakest part of this point of view [the iterative conception] is that the reasoning leading to the concept of a well-founded set [i.e. a set in the cumulative hierarchy] uses the well-ordering of the layers, and the notion of a well-ordering [. . .] is far from being simple or conceptually

[I]n order to state the axioms for a formal system, including all the types up to a given ordinal α, the notion of this ordinal α has to be presupposed as known, because it will appear explicitly in the axioms. On the other hand, a satisfactory definition of the transfinite ordinals can be obtained only in terms of the very system whose axioms are to be set up. (Gödel *1933: 8)

Gödel, however, noticed that we need not assume that the notion of an ordinal is already known to make sense of the cumulative hierarchy:

The first two or three types already suffice to define very large ordinals. So you can begin by setting up axioms for these first types, for which purpose no ordinal whatsoever is needed, then define a transfinite ordinal α in terms of these first few types and by means of it state the axioms for the system, including all classes of type less then α. (Call it S_α.) To the system S_α you can apply the same process again, i.e., take an ordinal β greater than α which can be defined in terms of the system S_α and by means of it state the axioms for the system S_β including all types less than β, and so on. (Gödel *1933: 8–9)

In Gödel's view, sets are what is obtained by iterating the *set of* operation as far as possible starting from some given objects. This, recall, gives rise to the cumulative hierarchy because the *set of* operation is 'substantially the same' as the powerset operation. What characterizes Gödel's account of the hierarchy is that it is *autonomous*, i.e. it does not presuppose an antecedently given system of ordinals along which the iteration takes place. Let us spell out the account in some detail.[7]

Gödel is working with the idea that we take the integers as individuals.[8] For generality, let us consider the case where we begin with no individuals. Hence, we have the empty set and the set containing the empty set, sets of those, sets of *those*, and so on, and then there is an infinite set, and sets of its elements, sets of those, sets of *those*, and so on. This characterization can be understood without presupposing a 'definition of the transfinite ordinals'; on the other hand, it tells us that we have the finite levels $V_0, V_1, V_2, \ldots$, and the levels $V_\omega, V_{\omega+1}, V_{\omega+2}, \ldots$. These levels, however, 'already suffice to define very large ordinals' along which we can continue the iteration of the *set of* operation without danger of circularity. That is, given, say, the level $V_{\omega+2}$, we can set up the system $S_{\omega+2}$ and define a large ordinal α_1, along which we can iterate the powerset operation so as to reach

---

fundamental'. And Lavine (1994: 148) writes that 'the iterative conception presupposes the notion of "transfinite iteration". In effect, the ordinal numbers are supposed to be given in advance'.

[7] See in particular Gödel *1933; *1951. Attempts to reconstruct Gödel's position can be found in Tait 2001: 87–96 and Koellner 2003: 21–29.

[8] Koellner (2003: 21–25), although taking the account in Gödel *1951 to be one in which the initial level of the hierarchy is the set of all integers, claims that in Gödel *1933 V_0 is the empty set. However, he offers no evidence for this claim. And in fact, Gödel's mention in Gödel *1933: 46 of the possibility of taking the integers as individuals, and his claim that the 'first two or three types' suffice to define very large ordinals, seem to indicate that the 1933 account and the 1951 account do not differ in this respect. Tait's (2001) reconstruction of Gödel's position is in agreement with our diagnosis.

level V_{α_1}. Since the existence of this ordinal was proved without assuming any prior conception of an ordinal, it can now be assumed without threat of circularity. This means that we can keep on going. That is, we can now set up another system S_{α_1}, in terms of which we can define an even larger transfinite ordinal α_2, which enables us to iterate the powerset operation so as to reach level V_{α_2}. And so on.

Gödel's strategy can be made a lot of sense of if we define ordinals not *à la* von Neumann (see Section 1.A) but by using a trick due to Scott (1955) and Tarski (1955). For, on the von Neumann definition, each ordinal α first appears in the hierarchy at V_α. As a result, one cannot define an ordinal β greater than α just by using resources available in the system including all levels less than α. Things change if we define ordinals by using the *Scott-Tarski trick*, however. This is a trick which is useful when we seem to be dealing with the set $\{x : \varphi(x)\}$ which does not exist. Rather than considering such a non-existent set, we consider the set of things which satisfy φ and occur at the lowest possible level in the hierarchy. Thus, in the case at hand, we define ordinals as order types of well-ordered sets and take the order type of a set a to be the set of all sets isomorphic to a which occur as low as possible in the hierarchy.[9] (Similarly, we can define the cardinality of a set a as the set of all sets equinumerous with a which occurs as low as possible in the hierarchy.) As a result, the Scott-Tarski ordinal $\omega + \omega$ will appear in the cumulative hierarchy well before we reach the level $V_{\omega+\omega}$. But once this ordinal is proved to exist, we can assume it without danger of circularity, and carry out more iterations of the powerset operation. But these iterations will, in turn, guarantee the existence of further ordinals, along which we can further iterate the powerset operation. Thus, Gödel's account can be naturally explained by using the Scott-Tarski definition of an ordinal, and has the resources to deal with the circularity objection.

But although this would be enough to dispense with the objection, there is a further point which deserves emphasis, and draws on the developments in set theory which we touched upon in Chapter 2. The axiomatization given by Scott (1974), of which SP is a descendant, shows that the worry that the notion of a well-ordering is needed to grasp the iterative conception is really just an idle concern. In particular, what Scott provided is an axiomatization of set theory which, albeit sanctioned by the iterative conception, does not assume a previous conception of the hierarchy as constituted by levels ordered by the ordinals. Rather, starting from certain ele-

---

[9] Note that in using the notion of the cumulative hierarchy in the Scott-Tarski definition of cardinals and ordinals, we have implicitly made use of the Axiom of Foundation. However, the Axiom of Foundation is, strictly speaking, not needed. All that is needed is that every set is the same size as a well-founded set. This is known as Coret's (1964) Axiom B and is provable in ZFC minus the Axiom of Foundation (and hence in the theories to be discussed in Chapter 7). See Forster 2003.

mentary facts about levels, which, as we have seen, he called 'partial universes', he established facts about sets and levels. Notably, what is assumed about the levels does not include that the levels are well-ordered. More specifically, he showed that, assuming the Axioms of Restriction and Accumulation, we can *prove*, together with the Axioms of Separation and Extensionality, that all axioms of Z except for Infinity hold, that every set is well-founded and, crucially, that the levels are well-ordered by membership. And whilst the Axiom of Accumulation might be taken to presuppose that the levels are partially ordered, it does not presuppose that they are *well*-ordered.

The upshot is that the fact that the levels of the hierarchy are well-ordered is not required to grasp the iterative conception, but is a consequence of it. I conclude that we do not need a prior and different notion of set to make sense of the notion of the cumulative hierarchy, and the circularity objection fails.

3.3 The No Semantics Objection

The third objection is to the effect that if we take all sets to be those in the hierarchy, we cannot provide a semantics for set theory. The objection, which we may call the *no semantics* objection, is summed up by Priest as follows:

The specification of the semantics of a first-order language is normally thought of as being couched in a set-theoretic metalanguage. Hence the language the specification of whose semantics poses problems is set theory itself. The interpretation of a set-theoretic language is a pair, $\langle \mathcal{D}, \mathcal{I} \rangle$, where $\mathcal{D}$ is the domain of all sets. Assuming the cumulative hierarchy to be the correct account of set, the semantics of set theory thus becomes impossible to specify in a coherent fashion. (Priest 2006b: 37)

Versions of the objection can also be found in Weir (1998a) and Priest (2002: 158–159). Let us spell it out in more detail.

The starting point is that we want to develop a model-theoretic semantics for a given language $\mathcal{L}$. The semantics ought then to deliver a definition of logical validity for $\mathcal{L}$ – that is, tell us what it is for an argument in that language to be logically valid, or for its conclusion to be a logical consequence of its premises. The standard way of doing this is by giving a Tarskian model theory for the language.

An interpretation is defined in the usual way, that is as a structure $\mathfrak{A}$ consisting of a domain $\mathcal{D}$ and an interpretation function $\mathcal{I}$ (see Section 1.1). Now let S be an assignment of values to variables, $S(x)$ the value of x under S, and an x-variant of S an assignment which assigns the same values to all variables except possibly x.

Then, via the standard recursive clauses, we define the notion of *truth in an interpretation* for the language in question:

(1) If S is an assignment of $\mathcal{D}$-elements to the variables, then $\mathcal{I}_S(x) = S(x)$. If a is an individual constant, $\mathcal{I}_S(a) = S(a)$.

(2) $\mathfrak{A}, S \models F(t_1, \ldots t_n)$ iff $\langle \mathcal{I}_S(t_1), \ldots \mathcal{I}_S(t_n) \rangle \in \mathcal{I}(F)$.

(3) $\mathfrak{A}, S \models \varphi \wedge \psi$ iff $\mathfrak{A}, S \models \varphi$ and $\mathfrak{A}, S \models \psi$.

(4) $\mathfrak{A}, S \models \varphi \vee \psi$ iff $\mathfrak{A}, S \models \varphi$ or $\mathfrak{A}, S \models \psi$.

(5) $\mathfrak{A}, S \models \neg\varphi$ iff $\mathfrak{A}, S \not\models \varphi$.

(6) $\mathfrak{A}, S \models \forall x\varphi$ iff, for each x-variant S' of S, $\mathfrak{A}, S' \models \varphi$.

(7) $\mathfrak{A}, S \models \exists x\varphi$ iff, for some x-variant S' of S, $\mathfrak{A}, S' \models \varphi$.

Logical validity for $\mathcal{L}$ is then defined in the obvious way in terms of truth in an interpretation: the argument from the set of sentences Σ to φ is *logically valid* ($\Sigma \models \varphi$) if and only if every interpretation making all members of Σ true makes φ true. A sentence φ is then *logically true* if and only if the argument with no premiss of which it is the conclusion is logically valid.

We are now in a position to state the no semantics objection more precisely. Consider the case of iterative set theory, which for present purposes will be our base theory Z^+. Since the set-theoretic quantifier is standardly taken as ranging over all sets, it seems that one of the interpretations quantified over in the definition of logical validity for $\mathcal{L}_\in$ – the *intended* interpretation – will have the set of all sets as its domain. But there can be no set of all sets in Z^+, on pain of contradiction. Hence, the objection goes, if we take all sets to be those in the hierarchy, we cannot give the usual model-theoretic definition of logical validity.

As in the case of the circularity objection, the problem is well-known to mathematical logicians and philosophers. As far as I am aware, it was first pointed out by Kreisel (1967), who put it as follows. Consider a sentence φ in $\mathcal{L}_\in$ with quantifiers ranging over all sets. According to the definition of validity, φ is logically true if and only if it is true in all interpretations whose domains are sets. However, since the intended interpretation is not among these interpretations, it seems that φ might be logically true and yet false in the cumulative hierarchy. But, the thought goes, we want our account of logical validity for $\mathcal{L}_\in$ to be such that if a sentence is logically true, then it is true in the hierarchy. Similarly, the problem in the case of arguments is that, according to the definition, an argument is valid just in case it preserves truth in an interpretation. But, since the intended interpretation is not among these interpretations, an argument involving sentences quantifying over all sets could be valid and yet fail to preserve truth.

Kreisel's way of putting the objection makes it clear that, contrary to what Priest seems to suggest (see, e.g., 2006b: 36–37), the problem does not lie in giving a definition of validity for $\mathcal{L}_\in$ in Z^+. As Priest himself (2006b: 277) admits, an account of validity defines the notion for an arbitrary language of a certain kind, and is usually given in a language such as $\mathcal{L}_\in$ by using resources available in Z^+.[10] But since $\mathcal{L}_\in$ is itself a language of the kind in question, we *can* give the required definition of validity in Z^+. What we *cannot* do in Z^+ is prove the existence of a model for the language of set theory and show that it is a model of Z^+. The problem, then, is not so much that we cannot give a definition of validity for $\mathcal{L}_\in$ in Z^+ – we can – but that we cannot show that the definition in question is extensionally correct, in the sense that, e.g., if a sentence is logically true it is true in the intended interpretation.

The issue is large and goes beyond the scope of this book. The best we can do is present solutions that the iterative theorist might adopt. It is important to stress, however, that the fate of the iterative conception might not turn on the viability of this solution for the following reason: it is far from clear that set theories based on conceptions of set other than the iterative one fare any better with respect to the no semantics objection.

For the objection can be raised just as much against the limitation of size conception and the graph conception, since both of these conceptions lead to set theories which do not admit of a universal set (see Chapters 5 and 7 respectively). Moreover, set theories based on the stratified conception (to be discussed in Chapter 6), whilst allowing for the existence of a universal set and being on the face of it more promising in this respect, must impose certain restrictions on the Axiom of Separation to avoid the set-theoretic paradoxes. In particular, if an interpretation assigns a subset of the domain as the interpretation of a monadic predicate, it would seem that some intuitive interpretations for monadic predicates will not be realized by any interpretation (see Rayo and Williamson 2003: 333).[11] On the other hand, the naïve set theories to be discussed in the next chapter allow for the existence of such a set and do not impose restrictions on the Axiom of Separation (since they allow for the full Comprehension Schema). This might suggest that the problem does not arise for them, and Priest certainly thinks that this is the case (see 2006b: 258–259; 277–278). His claim is that, in a set theory based on the naïve

---

[10] The definition of validity for a language such as $\mathcal{L}_\in$ can be carried out in the subsystem of PA_2 known as RCA_0 (see Simpson 1999: 23–26). Since PA_2 is known to be equivalent to Z minus the Powerset Axiom (see Burgess 2005: 65–72), it follows that Z^+ is more than enough to give the Tarskian definition of validity for $\mathcal{L}_\in$.

[11] This has been challenged by Schindler (2019: 424–425). If the semantics for iterative set theory can be carried out in theories like NF, this is good news for them as theories of objectified properties, which, we shall argue in Chapter 6, is how they are best understood.

conception we can define the notions of interpretation and truth in an interpretation (as in Z^+) for $\mathcal{L}_\in$ *and* prove the existence of the intended interpretation for the theory. However, the reasons he offers for thinking that are essentially that the model theory for his naïve set theory can be carried out in ZF using classical logic, and '[t]he paraconsistent logician can [...] simply appropriate the results' (2006b: 259). But this latter claim is based on the fact that he is taking for granted that one of the strategies he adopts for developing a naïve set theory – the model-theoretic strategy – is successful. But, as we shall see in the next chapter, this strategy fails. So can the model theory for naïve set theory be carried out within naïve set theory itself? This is an open question, and the weakness of naïve set theory, on both of the versions that we shall explore in the next chapter, means that the answer is by no means trivial. Thus, even if the no semantics objection were successful against the iterative conception, this would not immediately show that there are conceptions that fare any better in this respect.

Having said that, it would be desirable for the defender of the iterative conception to say something about how to deal with the no semantics objection. It is to this issue that we now turn.

3.4 Higher-Order Semantics

Many suggestions have been made to deal with the no semantics objection from the perspective of the defender of the iterative conception.

One strategy consists in giving up on the idea that the set-theoretic quantifier is to be interpreted as ranging over all sets (see, e.g., Parsons 1974; Lear 1977). According to this strategy, we never quantify over all sets, but only over the extension of 'set' at a certain time, which, in turn, depends on our conception of the set-theoretic hierarchy at a certain time. This understanding of set-theoretic quantification provides an easy way to deal with the no semantics objection. For, according to the proposal, our conception of the hierarchy is never complete and can always expand so that what at a certain time appears to be the universe of sets can always be later recognized as a set. And, on this view, this is exactly what happens when we give a semantics for set theory – or, better, for set theory as understood at a certain time. When we define validity for the language of set theory, we expand our conception of the hierarchy and recognize that what we thought was the universe of all sets really is a set, which can now act as the domain of the intended interpretation of the language being defined.[12] It goes without saying that this procedure can be

---

[12] The view described here is probably closer to Lear's. Parsons's view seems rather to be that the range of the set-theoretic quantifier is indeterminate between a number of possible interpretations given by interpretations whose domains, on a certain conception of the hierarchy, appear as the universe of sets. As our conception

iterated: we could, if we wanted, try and give a semantics for this new and more encompassing interpretation of set theory. If we did, we would realize that what we thought was the universe of all sets on the basis of this new interpretation is also a set. And so on.[13]

Some (e.g. Lear) have suggested, moreover, that once we take the extension of 'set' as changing over time in the way adumbrated, the semantics of set theory should be non-classical. For the purposes of this chapter, however, we can set the suggestion aside, since what we are interested in ascertaining here is whether there are any prospects for the idea that the defender of the iterative conception can make use of the classical Tarskian definition of validity whilst holding on to the idea that we can quantify over *all* sets.[14]

A popular strategy to achieve this has been to deny that the model theory for unrestricted languages is to be carried out by taking the domain of quantification to be a set. However, it is a familiar thought that not very much is solved *simply* by leaving things as they stand and taking the domain to be not a set but a different kind of object, a *proper class*. For on the one hand, we want $\mathcal{D}$ to be a member of other sets – for a start, it is a member of the ordered pair $\langle \mathcal{D}, \mathcal{I} \rangle$. On the other hand, proper classes are usually characterized as those collections which cannot be members. The familiar example here is von Neumann-Bernays-Gödel set theory (NBG), in which we have two types of objects: sets and classes. Every object is a class but some classes are not sets – they are *proper* classes, namely those classes that cannot be members of other classes. The theory can be finitely axiomatized using quantification over classes and is a conservative extension of ZFC.

We might drop the requirement that proper classes cannot be members (as is done in Ackermann 1956). But if we do, it is often objected, it looks as though we are constructing, on top of the cumulative hierarchy of sets, another hierarchy of objects which we are depriving of the title of sets without having earned the right to do so. Whatever the merits of this objection,[15] the appeal to proper classes does not

---

of the hierarchy expands, we recognize that some of these interpretations are sets, and rule out some of these possible interpretations.

[13] The view can somehow be traced back to Zermelo (1930: 1233), who seems to have conceived of the universe of set theory as an open-ended but well-ordered sequence of (intended) models, each of which is strictly larger than the preceding one. For any of these models – where a certain collection 'appears as an "ultrafinite non- or super-set"' – we can give a definition of validity for it where the intended model is included among the models quantified over by ascending to the next model, where the domain of the previous one is 'a perfectly good, valid set'. Zermelo's view of the matter, and its relation with the issue of whether we can quantify over all sets, is also touched upon in Incurvati 2016.

[14] For a discussion of the consequences of the adoption of a non-classical semantics in the context of various axiomatizations of set theory, see Incurvati 2008.

[15] One might wonder whether this objection can have any bite against a truly minimalist approach to the iterative conception. Now it is true that one could put forward a minimalist view which holds that all there is to a collection is to occur at one level or another of the cumulative hierarchy or occur in a hierarchy of further levels on top of the cumulative hierarchy. But, other things being equal, such a view will not be as natural

solve the problem but only postpones it: even in set theories where classes can be members of other sets, they cannot usually be members of themselves. This means that even if there can be a model with a domain consisting of all sets, there can be no model which contains all proper classes, and now we have the analogous problem that truth in all interpretations does not imply truth in the universe of all collections.[16]

The moral that some have been willing to draw is that we should completely give up on the idea that the domain of quantification is an object, and develop model theories accordingly. Earlier attempts in this direction were made by Boolos (1985), but the thought has more recently been developed by Rayo and Williamson (2003), building on work by Rayo and Uzquiano (1999). On their proposal, set-theoretic models are replaced by second-order models: rather than being identified with single set-theoretic objects, models are taken to be given by the objects which a monadic second-order variable I is true of. On this *higher-order semantics*, the 'domain' is therefore not construed as a set, but as whatever is in the range of our quantifiers: when we speak of the domain, we are not speaking of an object of some kind; rather we are speaking of the objects which are the values of I. By working within $ZFCU_2$, Rayo and Williamson go on to show that, by adopting this approach, we can define interpretations for ZFCU in which I ranges over all sets, and prove completeness and other meta-theoretic results for it. In other words, it is possible to carry out substantial meta-theoretic investigation for an unrestricted first-order language in its second-order counterpart.

But what is the significance of all this? If we accept the second-order theory in which the meta-theory is carried out, we can show that the definition of validity for the language of its first-order counterpart is extensionally correct. Thus, as long as one does not have qualms with second-order quantification, it is indeed possible,

---

or simple as the view according to which all there is to a collection is to occur at one level or another of the cumulative hierarchy. By inference to the best conception, the latter view is to be preferred. Of course, other things might not be equal, and in that case an account of the universe of collections which does make room for proper classes on top of the hierarchy might in the end be preferable. But that, it seems to me, is as it should be, and accounts very well for what the debate on proper classes has centred on, namely their usefulness.

[16] To be sure, some have suggested constructing proper classes as intimately connected with predicates – thus as logical collections – and in such a way that they are allowed to be members of themselves (see, e.g., Maddy 1983). This has two consequences. First, it makes the objection that proper classes are just sets in disguise seem much less compelling: proper classes look very much like objectified properties. Second, it opens up the possibility of the existence of proper class models which contain all proper classes. However, many proposals of this kind adopt an underlying non-classical logic. The prospects for developing a model theory in theories adopting logics of this sort are largely unexplored, and we can set this possibility aside for present purposes (see Leitgeb 2007 and Bacon 2013 for discussion of related issues). There are also proposals of this kind which *are* framed in a classical context (see Schindler 2019 and the idea of taking NF as a theory of objectified properties in Chapter 6). The issue in that case is whether they have the resources to develop model theory, both in terms of strength and in terms of the restrictions they place on the Axiom of Separation. See also fn. 11.

contra Priest, to specify the semantics of set theory in a 'coherent fashion'. To be sure, what has not been achieved is that a semantics for the language of first-order set theory has been given in the language of first-order set theory. Indeed, this *cannot* be achieved in a classical setting, as a suitable generalization of Tarski's theorem on the inexpressibility of truth shows (for details, see Florio 2014). Thus, as Rayo and Williamson themselves point out (2003: 353), to show that the definition of validity is extensionally adequate for the case of $ZFCU_2$, we would need to make use of resources available in third-order logic. More generally, Rayo and Williamson's strategy requires that to give the model theory for a language of order n we use a metalanguage of order $n + 1$. But if this is still found to be objectionable, then the no semantics objection seems to have become an objection against the use of classical languages in semantic theorizing, rather than on the possibility of giving a semantics for set theory.[17]

3.5 Kreisel's Squeezing Argument

Can the defender of the iterative conception respond to the no semantics objection and stick to a *set-theoretic* semantics, that is a semantics done using a set-theoretic meta-theory? The response we shall consider was suggested by Kreisel in the very same paper where he presented the problem with defining validity in an iterative set theory, and has been recently rehearsed by Field (2008: 46–48). Kreisel begins by considering three properties that a sentence φ might have. The first property is the property of being intuitively valid, expressed by the predicate Val. The second property is the property of being derivable in some preferred logical system, expressed by the predicate D. The third property is the property of being logically true according to the Tarskian definition of validity given above, expressed by the predicate V.

Kreisel sets out to show that V and Val are coextensive when φ is a first-order formula. His argument is as follows. First, we need to settle on some preferred first-order logic. It does not really matter what this logic is, as long as it has two features: first, it is the logic whose rules we accept as correct; second we have a completeness theorem for this logic relative to its Tarskian semantics. (The clauses of the semantics might then need to be modified if the chosen logic is not classical.)

---

[17] One should be careful not to overestimate the extent to which one needs to move to a language of higher-order in order to provide the required semantics. In particular, McGee (1997) has observed that in ZFC_2 one can define logical validity for arguments with finitely many premises using the second-order resources already available in ZFC_2 itself. In particular, let φ^U be the relativization of φ's quantifiers to U. Then we say that φ is a logical consequence of $\gamma_1, \ldots, \gamma_n$ iff the sentence $\forall U (\exists x U x \rightarrow (\gamma_1 [E / \in]^U, \ldots, \gamma_n [E / \in]^U) \rightarrow \varphi [E / \in]^U)$ holds.

Now say that our preferred logic is classical logic; the fact that we accept this logic as correct motivates the idea that if φ is classically derivable, $D_c(\varphi)$, then it is intuitively valid. That is:

$$D_c(\varphi) \rightarrow \text{Val}(\varphi). \tag{3.3}$$

The next step requires a little bit more reflection on the meanings of Val and V. Val expresses the property of being logically valid, and Kreisel takes this to mean *truth in all structures*. Now recall that, according to the Tarskian definition of validity, a sentence is logically true if and only if it is true in all interpretations, which, as we pointed out, are particular kinds of structures, set-theoretic structures (ordered pairs). But, Kreisel claims, if we know that a sentence is true in all structures, *a fortiori* we know that it is true in all *set-theoretic* structures. Hence, we have the following:

$$\text{Val}(\varphi) \rightarrow V(\varphi). \tag{3.4}$$

Thus, in Kreisel's view, the informal notion of truth in all structures for a first-order formula lies somewhere between derivability in our preferred formal system and Tarskian validity in that system. This already establishes that we can use the existence of a derivation in a preferred formal system to discover that a sentence lies in the extension of Val and the existence of a countermodel in the model theory to establish that it does not. What more can be established about the extension of Val? Since we have a completeness theorem for first-order classical logic,[18] we also have that

$$V(\varphi) \rightarrow D_c(\varphi). \tag{3.5}$$

Thus we can *squeeze* together the two formal properties between which the informal property of being true in all structures lies: the properties of being true in all structures, being true in all set-theoretic structures and being classically derivable are coextensive for first-order sentences.

This squeezing argument makes use of the Completeness Theorem for classical first-order logic. This means that the argument cannot be easily extended to second-order logic, a point which is often made against Kreisel's response to the no semantics objection (see, e.g., Weir 1998a: 770). Although this is true, it does not detract from the fact that the squeezing argument, if successful, at least shows that

---

[18] Note, in particular, that the Completeness Theorem for classical first-order logic can be proved in WKL_0, the subsystem of PA_2 obtained by adding to RCA_0 Weak König's Lemma as an axiom (see Simpson 1999: 35–37 and 139–141). This means that the theorem is available in Z^+, and can therefore be used by the defender of the iterative conception in her reasoning.

the Tarskian account of validity is extensionally adequate for the case of first-order languages. Moreover, in discussions of the argument, one point is often overlooked, namely that, if Kreisel is right, the notion of truth in all structures for a second-order formula, just like the analogous notion for a first-order formula, lies somewhere between derivability in a second-order system and truth in all set-theoretic models.

This is because, according to Kreisel, (3.3) and (3.4) can be accepted in the case in which φ is second-order just as much as they can in the case in which it is first-order. For, in his view, even when φ is second-order, we can accept that if φ is derivable in a standard system of second-order classical logic, then it is intuitively valid, since this was the reason why we accepted the system in the first place. And, similarly, the fact that φ is second-order does not affect the fact that if φ holds in all structures, then it holds in all set-theoretic structures. The latter implication, in particular, means that although there may be some second-order formulae that are not true in the hierarchy but happen to be true in all set-theoretic models, we can still make extensive use of Tarski's definition of validity in the case of second-order languages to establish things in the intended model.

Weir, however, thinks that there is a major problem with the squeezing argument. He writes:

Consider Kreisel's informal 'proof' that $[\text{Val}(\varphi) \leftrightarrow V(\varphi)]$. In the right to left direction, the argument appeals to completeness for $[\models]$, and then the intuitive soundness of the classical proof system. This last, however, is something a proponent of naïve set theory is bound to contest. Such a proponent will look at things in the following way: we had a package consisting of a set of classical operational inference rules, a set of classical structural rules and a naïve rule axiom scheme for the notion of class. It was discovered by Russell, Zermelo, Burali-Forti and others that the package is a trivially inconsistent one, yielding all sentences as consequences. The standard response is to abandon the last element, but the 'ingénue' insists we must abandon one of the first two [...], that is we must reject the soundness of the standard notion of proof. The Kreisel argument completely begs the question if directed against someone seriously considering as a response to the paradoxes and the semantic problem that one alter the logic not the set theory. (Weir 1998a: 772)

But who is begging the question here? The iterative theorist was challenged to provide a reason to regard the Tarskian definition of validity as extensionally adequate. To meet the challenge, she has offered an argument based on the assumption that the rules of the proof system she accepts are intuitively sound *for her*. To reply that she cannot make this assumption because she is addressing worries raised by the naïve theorist ignores the dialectical context. For the squeezing argument is not intended to persuade someone who rejects classical logic to accept that the Tarskian definition of validity is extensionally correct; rather the argument is intended to show that someone who accepts classical logic, for the very reason why she accepts it, has reasons to regard the Tarskian definition as extensionally correct. Someone who does not think classical logic is in order will, to be sure, reject the assumption that the classical rules are intuitively sound; but this does not detract from the fact that the squeezing argument can provide the defender of the iterative conception with a reason for thinking that Val is coextensive with V.

3.6 The Status of Replacement

As mentioned in the previous chapter, the status of Replacement on the iterative conception is controversial. Boolos (1971: 26–27; 1989: 97) famously denied that the axiom is justified on the iterative conception and so did others (see, e.g., Potter 2004: esp. 221–231). We now review and discuss the main arguments offered in the literature to the effect that the iterative conception sanctions Replacement.

Gödel's Argument

The first argument is due to Gödel, and is reported by Wang as follows:

From the very idea of the iterative concept of set it follows that if an ordinal number α has been obtained, the operation of power set $\mathcal{P}$ iterated α times leads to a set $\mathcal{P}^\alpha(\emptyset)$. But, for the same reason, it would seem to follow that if, instead of $\mathcal{P}$, one takes some larger jump in the hierarchy of types, e.g. the transition $\mathfrak{R}$ from x to $\mathcal{P}^{|x|}(x)$ (where $|x|$ is the smallest ordinal of the well-orderings of x), $\mathfrak{R}^\alpha(\emptyset)$ is likewise a set. Now, to assume this for any conceivable jump operation [...] is equivalent to the axiom. (Wang 1974: 186)

We can shed some light on this passage by looking again at Gödel's unpublished papers on the foundations of mathematics and at his account of the iterative conception, which we touched upon in Section 3.2. Recall that in Gödel's view, sets are what is obtained by iterating the *set of* operation ('substantially the same' as the powerset operation) as far as possible starting from the empty set.

As before, we initially have the levels $V_0, V_1, V_2, \ldots, V_\omega, V_{\omega+1}, V_{\omega+2}, \ldots$ We are now in a position to define an ordinal α_1, along which we can iterate the powerset operation so as to reach level V_{α_1}. Now we have to consider 'some larger jump in the hierarchy of types', that is the operation that given a well-ordering of x allows us to iterate the operation of powerset $|x|$-many times over x, that is the transition $\mathfrak{R}$ from x to $\mathcal{P}^{|x|}(x)$. Elsewhere, Gödel puts the matter as follows:

But are we to an end now? By no means. For we have now a new operation of forming sets, namely, forming a set out of some initial set A and some well-ordering B by applying the operation 'set of' to A as many times as the well-ordered set B indicates. And setting

B equal to some well-ordering of A, now we can iterate this new operation again, which we can treat in the same way, and so on. (Gödel *1951: 306)

Notice that although Gödel is talking of new operations here, what we are really asked to consider is the powerset operation in a general context: given any set A and a well-ordering α, we can iterate the powerset operation α many times starting from A. If we work with the von Neumann definition of ordinal, we might now want an assurance that any given set will have a well-order: this would enable us to obtain long well-orderings that will take us quite far in the iteration process. This assurance is given by the Axiom of Choice, and for this reason Tait (2001: 90–91) and Koellner (2003: 23) have argued that Gödel is assuming the axiom in his discussion.[19] Let us see how the account unfolds under this interpretation. Gödel initially talks of an operation for forming sets where we simply have an initial set A and some well-ordering B; he then goes on to suggest a strategy for continuing the iteration based on 'setting B equal to some well-ordering of A'. This means that – granting AC and assuming that Gödel, when talking of an ordinal being 'definable' meant 'definable with parameters', as Koellner (2003: 23) claims we should – the iteration process can continue as follows. Given the parameter $V_{\omega+1}$, we can get a well-ordering of length $\beth_1 + 1$, which gives us a level $V_{\beth_1+1}$.[20] Given this level as a parameter, we can then get a well-ordering of length $\beth_{\beth_1+1}$, which gives us a level $V_{\beth_{\beth_1+1}}$, and so on. If we carry out this operation ω-many times, we reach the first $\beth$-fixed point, i.e. the least κ such that $\kappa = \beth_\kappa$ (see Koellner 2003: 23 for details).

As noted in Section 3.2, however, Gödel's account can be easily made sense of if we work with the Scott-Tarski definition of ordinal. Gödel's argument is based on the idea that the powerset operation should be iterated as far as possible, which seems to harmonize well with the iterative conception. If this is true, then, we

---

[19] Tait (2001: 91) and Koellner (2003: 23) also mention the fact that Gödel (*1933: 50) says that one of the weak spots in the theory of sets 'is connected with the axiom of choice'. However, it is far from clear that this constitutes evidence that Gödel is assuming AC in his account of the iterative conception. For the context in which Gödel is raising the problem of AC is one in which he is discussing 'the problem of giving a justification for our axioms and rules of inference' (*1933: 49). But the theory of sets he is considering is NBG, which includes AC. This might be the reason why he mentions the axiom and the problems it raises. In fact, having said that one of the weak spots of the theory of sets is connected with AC, he immediately goes on to say that he does not want to go into the details of the problem 'because it is of less importance for the development of mathematics' (*1933: 50). This statement seems hard to explain if Gödel took AC to be crucial to his account of the iterative conception, which is supposed to sanction the theory of sets, including those parts of it needed for the development of mathematics.

[20] The *beth function* $\beth_\alpha$ is defined by transfinite recursion as follows:

$$\beth_0 = \aleph_0;$$
$$\beth_{\alpha+1} = 2^{\beth_\alpha};$$
$$\beth_\lambda = \bigcup_{\alpha<\lambda} \beth_\alpha \text{ if } \lambda \text{ is a limit ordinal.}$$

can go further than Z^+ licences us to. For in Z^+, we can prove the existence of various Scott-Tarski ordinals greater than $\omega + \omega$, but we cannot prove that the powerset operation can be iterated along these ordinals. It therefore looks as though we are not iterating the powerset operation 'as far as possible'. To allow for this possibility we can add the following axiom, which is effectively a single instance of the Replacement Schema:[21]

> **Axiom 3.6.1** (Axiom of Ordinals). *For each ordinal α there is a corresponding level V_α.*[22]

Call Zf^+ the theory obtained by adding the Axiom of Ordinals to Z^+. In this theory we can further carry out the iteration of the powerset operation and prove the existence of further levels of the cumulative hierarchy. Indeed, the models of Zf_2^+ are precisely the V_α with α a fixed point of the $\beth$-operation. To this extent, the suggestion is in line with Koellner's and Tait's reconstruction of Gödel's account, according to which, as we have seen, we can iterate the powerset operation up to the first $\beth$-fixed point. On the other hand, we see that by expressing the idea that the iterative process has to be carried out as far as possible with the Axiom of Ordinals, we can make do without AC and without taking definability to be definability with parameters, thus avoiding the main shortcoming of Koellner's and Tait's reconstruction of Gödel's account. So for instance, the existence of a level ω_1 follows from the Axiom of Ordinals and the fact that Z^+ proves the existence of an ordinal ω_1. More generally, since Hartogs's Theorem[23] already guarantees the existence of long well-orderings, if the Axiom of Ordinals is justified on the iterative conception, we do not need AC to go as far in the iteration process as Gödel is asking us to.

At this point, Gödel asks us to continue the iteration process by introducing new operations. How are we to do that? Gödel suggests that we *generalize* from the previous procedure:

So the next step will be to require that *any* operation producing sets out of sets can be iterated up to any ordinal number. (Gödel *1951: 306)

---

[21] This is in line with Parsons's (1977: 513, fn. 14) conclusion that 'sets and "stages" ought to be "formed" together, so that the formation of certain sets should make possible going on to further stages. However, the most obvious principle of this kind, that if a well-ordering has been constructed then there is a stage such that the earlier stages are ordered isomorphically to the given well-ordering, is weaker than the axiom of replacement'.

[22] As the discussion makes clear, the axiom is to be interpreted on the basis of the Scott-Tarski definition of ordinal. On the von Neumann definition, the axiom is trivial, since each von Neumann ordinal α first appears in the hierarchy at V_α. This is why the von Neumann treatment of ordinals is not available within Z^+, but is available once we add to it the Axiom of Ordinals.

[23] Hartogs's Theorem, provable in Z^+, says that there is no cardinal κ such that, for every ordinal α, κ is greater than or equal to the cardinality of α.

It is difficult, however, to see why the idea of iterating any operation along any given well-ordering should gain any support from the iterative conception of set. It may be taken to fit well with the iterative conception that the powerset operation is to be iterated as far as possible. However, it is not entirely clear why this should be the case with respect to *any* operation. Why is it that carrying out one of these operations as far as possible does not lead us outside the hierarchy? Gödel's argument, albeit showing that the idea of iterating the *set of* operation as far as possible sanctions the Axiom of Ordinals, fails to show that the idea sanctions full Replacement.

Nonetheless, the fact that the argument lends support to the Axiom of Ordinals should not be underestimated. For the standard examples of theorems of ordinary mathematics which are not provable in Z^+, most notably Borel determinacy (see Friedman 1971), are provable in Zf^+ (see Martin 1975). Moreover, the Axiom of Ordinals guarantees that the recursive definition of the levels of the hierarchy as a sequence of V_α is well-defined for every α. This latter fact is important because, together with some well-known facts about Z^+, it shows that the consequences of Replacement mentioned by Boolos in support of its fruitfulness are already consequences of Zf^+:

in addition to theorems about the iterative conception, the consequences of replacement include a satisfactory, if not ideal theory of infinite numbers, and a highly desirable result that justifies inductive definitions on well-founded relations. (Boolos 1971: 27)

The Argument from Cofinality

The next argument is probably the standard argument to the effect that Replacement is justified on the iterative conception. An early statement of the argument can be found in Shoenfield 1965: 239–240. The argument is also given in Wang 1974: 186, and is expanded upon in Shoenfield 1977: 323–326, which will be the focus of our discussion.

Shoenfield begins by laying down a possible guide for determining the existence of stages:

[T]he fundamental question for us is: given a collection **S** of stages, is there a stage after all of the members of **S**? [...] A possible answer to our fundamental question is this: there is a stage after all the stages in **S** provided that we can imagine a situation in which all of the stages in **S** have been completed. [...] At best, this is a very vague answer; for it is not at all clear in general what we can or cannot imagine. It does, however, provide a useful guide for obtaining more precise principles on which we can base axioms. Specifically, there are three cases in which our vague answer leads us to conclude that there is a stage after each member of **S**. The first is the case in which **S** consists of a single stage. The second is that in which **S** consists of an infinite sequence $S_0, S_1, \ldots$ of stages. The third case is that in which we have a set x and a stage S_y for each y in x, and **S** consists of the stages S for y in x. (Shoenfield 1977: 323–324)

There are several issues surrounding Shoenfield's talk of imagination in this passage, some of which are connected with the problems besetting the constructivist approach to the iterative conception (see Section 2.3). For present purposes, however, let us focus on Shoenfield's principles for determining the existence of stages.

To suppose that there is a stage after all of the members of **S** in the first two cases gives rise, respectively, to the usual operations of taking powerset and summing up a transfinite sequence of levels. To suppose that there is such a stage in the third case gives rise to the so-called *principle of cofinality* (see also Boolos 1971: 26):

$$Cof \quad \forall x \exists s \forall y (y \in x \rightarrow \exists t (t < s \wedge yFt)).$$

It is easy to see that this principle delivers Replacement. For recall that what Replacement says is that if f is a function and a is a set, then there is a set of all $f(b)$s for $b \in a$. So, for each $f(b)$, we just let $S_{f(b)}$ be a stage at which $f(b)$ is formed. This will also be a stage at which b is formed, since, according to the account of the iterative conception offered by Shoenfield and Boolos, each set is formed at all stages that are later than any one at which the set is formed (see, e.g., Boolos 1989: 88). But then, by the cofinality principle, we can conclude that there is a stage after all the stages $S_{f(b)}$s for $b \in a$. The desired set of all $f(b)$s is then formed at this stage.

However, we now need an argument to the effect that the principle of cofinality is true on the iterative conception. Shoenfield writes:

Suppose that as each stage S is completed, we take each y in x which is formed at S and complete the stage S_y. When we reach the stage at which x is formed, we will have formed each y in x and hence completed each stage S_y in **S**. (Shoenfield 1977: 324)

This argument can be easily turned into an explicit argument for Replacement, as follows.[24] We know that each y in x is formed at some stage S preceding the stage at which x is formed (since we know that x is a set). Moreover, we know that for each y, there is a stage at which each $f(y)$ is formed (since we know that if y is a set, $f(y)$ is a set too). What we want is a stage before which each $f(y)$ is formed.

---

[24] In fact, this is how the argument is presented by Wang (1974: 186), who phrases it in constructivist terms: 'Once we adopt the viewpoint that we can in an idealized sense run through all members of a given set, the justification of [the Axiom of Replacement] is immediate. That is, if, for each element of the set, we put some other given object there, we are able to run through the resulting multitude as well. In this manner, we are justified in forming new sets by arbitrary replacements. If, however, one does not have this idea of running through all members of a given set, the justification of the replacement axiom is more complex'.

According to Shoenfield, we can see that there is such a stage by going through the stages preceding the stage at which x is formed: each time we reach a stage S at which one of the ys is formed, we go through the stages necessary to form the corresponding $f(y)$. By the time we have reached the stage at which x is formed, all the $f(y)$s will have been formed, which guarantees that there is a stage at which they are collected into a set.

Note, however, that the argument seems to assume that if a process can be completed, and we replace each stage of the process with a process that can be completed, then the maxi-process consisting of all these processes can itself be completed. But this is just the Axiom of Replacement reformulated in terms of stages and processes.[25]

Thus, it is not clear that the principle of cofinality follows from the iterative conception of set. Boolos concluded that the principle

is an attractive further thought about the interrelation of sets and stages, but it does seem to us to be a *further* thought, and not one that can be said to have been meant in the rough description of the iterative conception. (Boolos 1971: 26)

However, even if one sides with Boolos on this, the cofinality principle certainly seems to harmonize well with the iterative conception, and can perhaps be seen as one way of spelling out the idea that the cumulative process through which the hierarchy is obtained should be iterated as far as possible.

The Argument from Reflection

The third argument for Replacement starts from the idea that what it means for the hierarchy to have as many levels as possible is for it to be *absolutely infinite*. The notion of absolute infinity goes back to Cantor, who attached great heuristic importance to it (see Hallett 1984: ch. 1; Jané 1995). The notion is not completely transparent, but a central feature of it, in Cantor's view, is that absolute infinity resists mathematical determination (see Hallett 1984: 13). Focusing on this feature suggests equating the hierarchy's absolute infinity with its not being uniquely characterizable by any property of a certain type $\mathcal{K}$. This, in turn, is taken to give rise to the following informal principle:

Reflection Principle. Let P be a property of type $\mathcal{K}$. If **V** has P, then there exists some V_α which has P.[26]

---

[25] A similar point is made in Parsons 1977: 513–514, fn. 15, although Parsons is there discussing the earlier Shoenfield 1965.

[26] Although Cantor never explicitly formulated it, the Reflection Principle has been argued to be faithful to his original ideas on absolute infinity by Reinhardt (1974: 190–192), Hallett (1984: 117–118) and Welch and Horsten (2016) among others.

Obviously, the class $\mathcal{K}$ cannot contain trivial properties such as $x = V$, lest a contradiction be easily obtained from the Reflection Principle (see Reinhardt 1974: 190). Little has been said, however, by way of a positive characterization of $\mathcal{K}$. Gödel took the Reflection Principle to hold when $\mathcal{K}$ is the class of *structural* properties (see Wang 1996: 283–285), but a full-fledged account of the notion of a structural property remains wanting.

The Reflection Principle is sometimes said to express the idea that the hierarchy cannot be reached from below (see, e.g., Paseau 2007: 33; Koellner 2009: 208). However, this idea seems better expressed by a more specific principle saying that V cannot be obtained through the application of any collection of (non-trivial) closure principles, such as closure under powersets:[27]

> **Closure Principle.** Let C be a collection of closure principles and let P be the property of being closed under C. If V has P, then there exists some V_α which has P.

A version of the Closure Principle is enunciated by Gödel in a passage reported by Wang (Wang 1974: 189; see also Wang 1996: 280). On the assumption that the class $\mathcal{K}$ of properties to which the Reflection Principle applies includes any property of the type *being closed under closure principles* C, the Reflection Principle implies the Closure Principle. The converse, however, need not be true, since $\mathcal{K}$ may contain other types of properties. Indeed, the Reflection Principle may be considered a generalization of the Closure Principle (see, e.g., Wang 1974: 189, reporting Gödel).

The Closure Principle, and hence the Reflection Principle, can be used to argue informally for the existence of large cardinals. Suppose we have come to accept that V is closed under powersets and unions of set-indexed families of sets (including uncountable ones).[28] By the Closure Principle, we can conclude that there exists a level of the hierarchy which is equally closed. Such a level has inaccessible height.[29] Similar arguments can be used to argue for the existence of larger small large cardinals,[30] such as Mahlos[31] (see Reinhardt 1974: 190–191 and Wang 1996: 280, reporting Gödel).

---

[27] The restriction to non-trivial closure principles is needed since the hierarchy is closed under the *set of* operation.

[28] Note that, formally, V being closed under unions of set-indexed families of sets is guaranteed by the Axiom of Union *in conjunction with* the Replacement Schema. Thus, for instance, Z does not prove the existence of a set $\bigcup_n \mathcal{P}^n(\omega)$.

[29] A cardinal κ is *inaccessible* if, as well as being uncountable, it is regular and strong limit. The former condition is equivalent to the sets below κ being closed under unions of set-indexed families of sets, the latter to their being closed under powersets.

[30] A large cardinal κ is said to be *small* if it is consistent that the axiom asserting the existence of κ holds in L (provided, of course, that it is consistent that it holds in V). Otherwise, κ is said to be *large*.

[31] A cardinal κ is *Mahlo* if it is regular and the set of inaccessibles below κ is stationary in κ, where a set $A \subseteq \kappa$ is *stationary* in κ if $A \cap C \neq \emptyset$ for each closed unbounded subset C of κ.

We must now turn the Reflection Principle into an axiom schema. In axiomatic set theory, properties are expressed by formulae of a formal language. We indicate with $\mathcal{L}_{\in}^n$ the language obtained by adding $\in$ to the language of nth-order logic. If $n = 1$, we simply omit the superscript (in line with our notation so far).

Next, we need a way to say that V_α has the property expressed by φ. We do this by taking φ to be relativized to V_α in the standard way, i.e. by restricting φ's quantifiers to $x \in V_\alpha$. We write φ^α to denote φ's relativization to V_α.

Formulae may contain parameters, which must also be relativized. If A is a parameter, we let A^α be its relativization to V_α, which is defined as follows. If A is a first-order parameter, $A^\alpha = A$. If A is a second-order parameter, A^α is $A \cap V_\alpha$. If A is a parameter of order $n + 1$ (with $n > 1$), A^α is $\{B^\alpha | B \in A\}$, where B ranges over nth-order classes.

We can now formulate a template for obtaining reflection principles for languages and parameters of varying complexity:

> **Reflection Principle Template** (RP). Let $\varphi(A_1, \ldots, A_k)$ be an $\mathcal{L}_{\in}^m$-formula with parameters $A_1, \ldots, A_k$ of order at most n. Then the following holds:
>
> $$\varphi(A_1, \ldots, A_k) \to \exists \alpha \varphi^\alpha(A_1^\alpha, \ldots, A_k^\alpha).$$

Actual axiom schemata can be obtained from RP by specifying the language in which φ is cast and the order of the parameters that are allowed. We use $\mathsf{RP}_{m,n}$ to denote reflection on $\mathcal{L}_{\in}^m$-formulae with parameters of order at most n. The strength of the Reflection Principle depends on the properties that it is taken to apply to. As one might expect, then, the choice of language and parameters has a severe impact on the strength afforded by RP.

The study of reflection principles began with Richard Montague (1957; 1961), who isolated the following principle and showed it to be a theorem of ZF:

> **Complete Reflection Principle** (CRP) (Montague). Let $\varphi(x_1, \ldots, x_n)$ be an $\mathcal{L}_{\in}$-formula. Then for each α, there is a $\beta > \alpha$ such that, for any $A_1, \ldots, A_n$ in V_β,
>
> $$\varphi(A_1, \ldots, A_n) \leftrightarrow \varphi^\beta(A_1, \ldots, A_n).$$

This asserts that any first-order property is *completely* reflected, i.e. reflected in *arbitrarily high* levels of the cumulative hierarchy. On the other hand, $\mathsf{RP}_{1,1}$ simply asserts that any such property is reflected at *some* level of **V**. This difference affects the strength of the two principles. For, as Azriel Lévy (1960) went on to prove, CRP is actually equivalent to Infinity and Replacement in the presence of the other ZF axioms (indeed, in the sole presence of the axioms of Extensionality, Foundation and Separation). If CRP is justified on the iterative conception, this would provide a possible route to the justification of these axioms (see, e.g., Paseau 2007). By contrast, whilst $\mathsf{RP}_{1,1}$ does afford the derivation of Infinity, it fails to deliver Replacement (see Lévy and Vaught 1961).[32] Thus, if RP is to be used to obtain Replacement, we need to allow higher-order parameters.

Let us therefore consider $RP_{1,2}$. As well as Infinity, the Replacement Schema is now derivable. This follows from the fact that $RP_{1,2}$ implies CRP. In fact, many of the ZF axioms are now derivable. In particular, the theory consisting of Extensionality, Foundation, Separation and $RP_{1,2}$ implies ZF.

We have now reached the limit of $RP_{1,2}$, since the theory in question is in fact *equivalent* to ZF (see Gloede 1976: 293). If we want to generate axiom schemata which non-conservatively extend ZFC in the context of a first-order language, the natural move is to consider $RP_{1,3}$. The following result, however, tells us that this would be inadvisable:[33]

Theorem 3.6.2. $RP_{1,3}$ *is inconsistent.*

Tait (1998) has suggested repairing the situation by foregoing certain negated formulae. Koellner (2009) has shown that $RP_{3,4}$ with the restriction suggested by Tait is still inconsistent.

So let us consider instead the language $\mathcal{L}^2_\in$. Bernays (1961; 1976) showed that the theory consisting of Extensionality, Second-Order Separation, Foundation and $RP_{2,2}$ implies ZF_2 (ZF with the Replacement Schema replaced by the corresponding second-order axiom). What is more, the theory also implies the existence of inaccessibles, Mahlos, weakly compacts[34] and other small large cardinals.

This latter fact can be proved by *reflecting on reflection*, starting from the fact that $RP_{1,2}$ holds in the universe. We could not do this in ZF because the fact that the hierarchy satisfies $RP_{1,2}$ is not expressible by a finite number of $\mathcal{L}_\in$-sentences. However, there is a single $\mathcal{L}^2_\in$-sentence saying, in effect, that $RP_{1,2}$ holds:

$$\forall A \exists \alpha \langle V_\alpha, \in, A^\alpha \rangle \preceq \langle \mathbf{V}, \in, A \rangle,$$

where A is second-order and $\mathcal{M} \preceq \mathcal{N}$ means that $\mathcal{M}$ is an elementary substructure of $\mathcal{N}$. This sentence is provable in ZF_2, so by $RP_{2,2}$ we can conclude that it already holds in some V_κ. Thus, we have:

$$\forall A \subseteq V_\kappa \exists \alpha < \kappa \langle V_\alpha, \in, A^\alpha \rangle \preceq \langle V_\kappa, \in, A \rangle.$$

---

[32] It does deliver more than Infinity, however. For instance, it enables one to prove Axiom (ρ), which is independent of Z.

[33] The result is noted by Reinhardt (1974: 196), but a proof seems to have first appeared in print only in Tait 1998: 481.

[34] Consider the language $\mathcal{L}_{\kappa,\kappa}$, which, besides using all formation rules of a finitary logic, allows the formation of up to κ long sequences of conjunctions, disjunctions, universal and existential quantifications. Then κ is *weakly compact* if $\kappa > \omega$ and, for any set Φ of sentences in $\mathcal{L}_{\kappa,\kappa}$ using at most κ non-logical symbols, if every subset of Φ of cardinality less than κ is satisfiable, then Φ is satisfiable.

By a result of Lévy (1960), this is equivalent to κ being inaccessible (see Kanamori 2003: §6). Thus, the sentence '$RP_{1,2}$ holds and there is an inaccessible cardinal' is now provable. By $RP_{2,2}$, we conclude that there are two inaccessibles. Carrying out this reasoning over and over again, we arrive at the conclusion that there is a proper class of inaccessible cardinals. Similar arguments can be used to prove the existence of Mahlo cardinals: the informal arguments for the existence of inaccessibles and Mahlos can be captured with the help of RP together with the expressive resources of second-order languages.

We mentioned earlier that Koellner (2009) has shown that, even when restricted to formulae of a certain kind, $RP_{3,4}$ is inconsistent. In the same paper, he offers an array of results which he takes to show, more generally, that principles of the form RP, if consistent, are weak, in that they cannot even yield the Erdös cardinal $\kappa(\omega)$,[35] and hence cannot lead to what Koellner describes as a 'significant reduction in incompleteness'.[36]

Koellner (2009: 217–218) concludes by expressing pessimism about the possibility of justifying large cardinal axioms *in general* on the basis of the Reflection Principle. The issue remains open, however, and indeed some set theorists have claimed that the Reflection Principle, and the underlying idea that the hierarchy is absolutely infinite, can be used to motivate the existence of large large cardinals via elementary embeddings. For remarks along these lines, see Martin and Steel 1989: 72–73. A systematic defence of this position can be found in Welch and Horsten 2016 and Welch 2014.

It is often claimed that the absoluteness of the hierarchy is part of the iterative conception. Wang offers the following argument for the idea that the hierarchy cannot be characterized by any non-trivial property:

The power and justification of the reflection principle depend on the following consideration. For our knowledge, the 'iterative' process means that we can use all conceivable means to continue the process, that the process is 'unlimited' [...] We cannot see exactly how we can possibly reach V beginning with the null set according to the rank hierarchy; there are a lot of gaps in between. Therefore, we can never capture the full content of V and

---

[35] For every limit ordinal α, the *Erdös cardinal* $\kappa(\alpha)$ is the least κ such that for every partition of the finite subsets of κ into two cells, there is a subset of κ of order type α all of whose finite subsets lie in the same cell.

[36] Koellner (2009: 208) argues that a reasonable target for a significant reduction in incompleteness is the axiom 'For every set X, $X^\sharp$ exists'. Given this target, his result shows that principles of the form RP, if consistent, are *very* weak, since it implies that they cannot even deliver $0^\sharp$. This, if it exists, is the set of all formulae $\varphi(v_1, \ldots, v_n)$ of $\mathcal{L}_\in$ such that $\mathbf{L} \models \varphi(\kappa_0, \ldots, \kappa_n)$ for any strictly increasing sequence $\kappa_0, \ldots, \kappa_n$ of uncountable cardinals, and can be regarded as a subset of ω via Gödel numbering. The existence of $0^\sharp$ is equivalent to the existence of a non-trivial elementary embedding of $\mathbf{L}$ into itself. For any set A, $A^\sharp$, if it exists, is obtained by replacing $\mathbf{L}$ with $\mathbf{L}[A]$ (the smallest inner model $\mathcal{M}$ such that for every $X \in \mathcal{M}, A \cap X \in \mathcal{M}$) in the definition of $0^\sharp$.

any property which characterizes V uniquely can only do this 'by default' and does not tell us much about the elements of V. (Wang 1977: 319)

However, the argument just *assumes* that the iterative process is unlimited. Similarly, Paseau (2007: 33) simply states that 'Reflection and the associated idea that the hierarchy is absolutely infinite are arguably part and parcel of the iterative conception' (see also Martin 1998; 2001; Tait 1998).

Indeed, doubts are sometimes raised about whether this is the case (see, e.g., Potter 2004: 222–223): if we understand the iterative conception as maintaining that sets are the objects obtained by iterating the *set of* operation, then the idea that the hierarchy is absolute cannot quite be *subsumed* under the iterative conception.

Nonetheless, the idea that the *set of* operation is to be iterated as far as possible *harmonizes* well with the iterative conception: unlike any answer specifying some particular cardinality, it seems a principled and non-arbitrary answer to the question of how far the iteration process goes. And absolute infinity is a natural way of understanding the idea that the iteration process is to be carried out as far as possible, and it appears to deliver Replacement and various large cardinal axioms.

3.7 Conclusion

The defender of the iterative conception can provide satisfactory responses to the missing explanation objection and to the circularity objection. As for the no semantics objection, many of the standard strategies for dealing with it have been found wanting. However, the approach based on giving a semantics by making use of higher-order entities and the approach based on Kreisel's squeezing argument, I have argued, are both promising. Moreover, as I have pointed out, it is far from clear that the no semantics objection affects the iterative conception any more than the other conceptions of set.

As for the status of Replacement, even if the idea of iteration as far as possible is perhaps not *part* of the iterative conception, it harmonizes well with it. Understood or augmented in this way, the iterative conception sanctions the Axiom of Ordinals and possibly the Replacement Schema. And when added to Z^+, the Axiom of Ordinals already suffices to account for the standard cases of examples from ordinary mathematics which cannot be reconstructed in Z^+.

What remains to be investigated is how the iterative conception fares with respect to the other conceptions of set available. This task will occupy us in the remaining chapters.

4

The Naïve Conception

Only the most naïve of
questions are truly serious.

Milan Kundera (1984: 139)

In Chapter 3, I considered three objections to the claim that the hierarchy covers all sets, and argued that none of them shows that the account of set based on the cumulative hierarchy is unsatisfactory. According to the method of inference to the best conception, however, the claim that all sets are those in the hierarchy is to be defended by showing that the iterative conception fares better than its rivals with respect to certain desiderata on conceptions of set.

For this reason, our next task is to examine and evaluate the conceptions of set which are usually considered alternatives to the iterative conception. In this chapter, we look at attempts to rehabilitate the naïve conception by changing the logic of set theory.

4.1 Paraconsistency and Dialetheism

Recall that according to the naïve conception of set, to every predicate there corresponds a set, and this is usually taken as motivating classical naïve set theory, the theory whose logic is classical logic, and whose axioms are Extensionality and Naïve Comprehension. In the previous chapter, I cast doubts on the idea that the Comprehension Schema is as intuitive as is sometimes claimed, since the reason why people find it intuitive might be due to the fact that what they have in mind is a conception of sets as sets of individuals, or that they are conflating two different concepts, the concept of set and the concept of objectified property. Having said that, it can be granted, for the purposes of this chapter, that the naïve conception of set is natural – in the sense that it can be easily acquired without prior knowledge or experience of sets and that we find it plausible or at least conceivably true – and sanctions the two axioms of naïve set theory. This conception, as we have seen, gives rise to a theory which is classically inconsistent and therefore, in the presence of the classical rule of *ex falso quodlibet* ($\varphi, \neg\varphi \vdash \psi$, also called *Explosion*), trivial: it proves everything. The received view is that this shows that something is wrong with the underlying conception: it is not the case that every condition determines a set. Thus, according to the iterative conception, the only instances of Comprehension which hold are those that hold in the cumulative hierarchy.

However, a number of philosophers, most notably Graham Priest (esp. 2006b), have tried to rehabilitate the naïve conception by suggesting that what the paradoxes show is that something is wrong with the underlying *logic*: rather than abandoning the Naïve Comprehension Schema, we should abandon classical logic. In particular, Priest suggests resuscitating the naïve conception by developing naïve set theory in a logic in which contradictions such as Russell's can be derived but tolerated – theories where it can be proved that the Russell set both belongs and does not belong to itself but a contradiction does not entail everything.[1]

We will see below that this task is harder than it might seem at first sight. For the moment, we notice two facts concerning this sort of attempt. First, the resulting set theory will have to be cast in a *paraconsistent* logic – a logic in which Explosion fails. Second, if the axioms of the theory are taken to be true, and its rules of inference to be truth-preserving, then proponents of these attempts seem to be committed to *dialetheism*, the view that there are true contradictions, or dialetheias.

Having noted these two key aspects of Priest's proposal, we now need to review and evaluate a number of strategies for developing a paraconsistent set theory. First, however, we will examine two constraints which, according to Priest himself (2006b: 248), a paraconsistent set theory should satisfy.

---

[1] For reasons of space, I cannot discuss the option of developing naïve set theory in a logic in which contradictions such as Russell's cannot be derived, but it will be helpful to say something about why the prospects for doing so do not look bright either. One strategy here is to curtail the rules for the logical constants, e.g. the conditional. Whilst promising for constructing a consistent naïve theory of properties (see, e.g., Field 2008), this strategy faces serious difficulties when used to develop a consistent naïve theory of *sets*. As shown by Field et al. (2017), Extensionality is inconsistent with a certain way of capturing naïveté given a reasonably strong logic. Another strategy is to tinker with the rules governing the structure of derivations in a formal system. An early instance of this strategy is pursued by Grišin (1974), who presents a naïve theory in a logic without the structural Contraction Rule of sequent calculus. However, this theory trivializes in the presence of Extensionality. More recently, the strategy has been developed by Weir (1998b; 1999), who suggests adopting a logic which avoids triviality by placing constraints on the way inferences can be chained together. The resulting non-transitive logic is very weak, and Weir's attempt to develop a decent amount of contemporary set theory is subject to criticisms similar to those levelled in Section 4.5 against one of the strategies for developing a paraconsistent set theory. Ripley (2015) presents a naïve set theory in a different non-transitive setting. His system shares many of the features (and problems) of naïve set theories based on LP, to be described below.

4.2 Neither Weak nor Trivial

A paraconsistent set theory based on the naïve conception will consist of Extensionality and Comprehension, and an underlying paraconsistent logic. Comprehension, recall, is the following schema:

$$\exists y \forall x (x \in y \leftrightarrow \varphi(x)), \tag{Comp}$$

where y does not occur free in φ. In certain paraconsistent set theories, the defining predicate φ is allowed to contain y free. This feature has been used by Routley to argue that AC is a consequence of Comprehension, although the argument is controversial (see Routley 1980: 892–962; Priest and Routley 1989). Unless otherwise stated, we will focus on the standard formulation of Comprehension.

The choice of the logic, however, is delicate, and can give rise to a number of different set theories, depending also on whether or not we take the conditional in the axioms of Comprehension and Extensionality to be material (see Priest 2006b: 248). We shall distinguish between the resulting theories by making explicit in their name the logic in which they are cast. So, for instance, NLP will be naïve set theory in the logic LP, to be described below. At first sight, then, there *seems* to be a great deal of freedom left in the development of a paraconsistent set theory. Still, Priest acknowledges that there are two constraints which a set theory should satisfy.

The first constraint is that the theory should not allow one to prove too much, and in particular should not be trivial. This means that it is not enough to just drop *ex falso quodlibet* so that we cannot infer anything from the Russell contradiction, the reason being that there are arguments which generate triviality from the Comprehension Schema without using Explosion. For instance, the set-theoretic version of Curry's Paradox tells us that in the presence of Comprehension triviality ensues as long as we have in our logic the rule of *modus ponens* and the Contraction rule:[2]

$$\frac{\varphi \to (\varphi \to \psi)}{\varphi \to \psi} \tag{Contr}$$

The derivation is as follows:

1. $\exists y \forall x (x \in y \leftrightarrow (x \in x \to \varphi))$ By (Comp)
2. $\forall x (x \in a \leftrightarrow (x \in x \to \varphi))$ Supp
3. $(a \in a \leftrightarrow (a \in a \to \varphi))$
4. $(a \in a \to (a \in a \to \varphi))$ By simplification

---

[2] Not to be confused with the *structural* rule of Contraction of standard sequent calculus systems. To be sure, it is possible to retain *modus ponens* and Contraction whilst avoiding triviality by placing constraints in the way steps in a derivation are chained together, as is done by Weir (see fn. 1 above). Here, I am setting aside this option, as well as the desperate strategy of blocking the derivation by dropping very basic rules such as $\exists$-elimination.

5. $(a \in a \to \varphi)$ 4, by (Contr)
6. $a \in a$ 3, 5, by *modus ponens*
7. φ 5, 6, by *modus ponens*
8. φ 1, 2–7, by ∃-elimination

Still worse, there is a variant of the paradox which shows that even if we drop Contraction, triviality ensues from Comprehension in the presence of *modus ponens* and the so-called *Modus Ponens Axiom*:

$$(\varphi \wedge (\varphi \to \psi)) \to \psi. \tag{MPA}$$

The new derivation is as follows:[3]

1. $\exists y \forall x (x \in y \leftrightarrow (x \in x \to \varphi))$ By (Comp)
2. $\forall x (x \in a \leftrightarrow (x \in x \to \varphi))$ Supp
3. $(a \in a \leftrightarrow (a \in a \to \varphi))$
4. $(a \in a \wedge (a \in a \to \varphi)) \to \varphi$ By MPA
5. $(a \in a \wedge a \in a) \to \varphi$ 3, 4, by substitutivity of equivalents
6. $a \in a \to \varphi$ 5, idempotence of conjunction
7. $((a \in a \to \varphi) \to a \in a)$ 3, by simplification.
8. $a \in a$ 6, 7 by *modus ponens*
9. φ 6, 8, by *modus ponens*
10. φ 1, 2–9, by ∃-elimination

    The second constraint is that our set theory should not prove too little: we want to be able to carry out a reasonable amount of standard set theory within it. This is in keeping with the desideratum we laid down in Section 2.7 when introducing the method of inference to best conception. If it is too weak, set theory is deprived of the role, with which it has traditionally been conferred, of giving a foundation for mathematics and of providing a tool for understanding the infinite. We need to preserve at least certain crucial results of set theory. As Priest puts it:

We might not require everything; we might be prepared to write off various results concerning large cardinality, or peculiar consequences of the Axiom of Choice. But if we lose too much, set theory is voided of both its use and interest. (Priest 2006b: 248)

The second constraint makes it harder to satisfy the first one: it is easy enough to find a logic that avoids triviality in the presence of Comprehension by irredeemably crippling the proof theory; it is much harder to find a logic which avoids triviality and yet allows us to develop a coherent and substantial amount of mathematics in it.

---

[3] The original version of Curry's Paradox appeared in Curry 1942. Its strengthened version is due to Meyer et al. (1979). In many relevant logics, to be discussed below, the Modus Ponens Axiom delivers Contraction.

In the light of Curry's Paradox and its strengthened version, one might consider either dropping *modus ponens* or giving up Contraction and the Modus Ponens Axiom. And in fact, both strategies have been pursued, which means that, initially, two strategies for developing a paraconsistent set theory suggest themselves, which, following Priest (2006b: 249–255), we may call the *material strategy* and the *relevant strategy*. We shall tackle them in turn.

4.3 The Material Strategy

The Logic of Paradox

According to the material strategy, triviality is avoided by invalidating *modus ponens* and taking the conditional in the Axioms of Comprehension and Extensionality to be the material conditional. That is, we take $\varphi \to \psi$ to be defined as $\neg\varphi \lor \psi$, and devise our paraconsistent logic in such a way that, besides Explosion, the inference

$$\frac{\varphi \qquad \neg\varphi \lor \psi}{\psi}$$

usually known as *disjunctive syllogism*, also fails. The biconditional $\varphi \leftrightarrow \psi$ is then defined in the usual way, that is as $\varphi \to \psi \land \psi \to \varphi$.

According to Priest (2006b: 249), a 'simple and natural choice here' is his Logic of Paradox (LP), first introduced in Priest 1979. The motivation for the choice can be given as follows. According to the dialetheist, as we have seen, there are sentences φ such that $\varphi \land \neg\varphi$ is true, and the statement asserting that the Russell set belongs to itself is one of those. Now suppose you believe that a conjunction is true if and only if both of its conjuncts are true. This means that φ and $\neg\varphi$ are both true. And suppose you believe, further, that a sentence is true if and only if its negation is false. Then, sentences like φ will be both true and false: the mistake of the classical logician lies precisely in assuming that there cannot be sentences of this kind.

Hence, we want a logic in which a sentence can be both true and false; moreover, when a sentence *is* both true and false, it receives a designated value since, according to the dialetheist, we should assert a sentence if it is *at least* true. LP implements both these facts in that its set of values π is $\{\{1\}, \{0\}, \{0, 1\}\}$, and its set of designated values δ is $\{\{1\}, \{0, 1\}\}$. For ease of exposition and readability, we will define T, F and B (for True, False and Both) to be $\{1\}$, $\{0\}$ and $\{0, 1\}$ respectively.

How should connectives and other logical constants behave in LP? Again, the dialetheist's answer is to preserve as much as possible of classical logic whilst admitting the value B, and to use basic principles about the connections between truth and falsity to decide the behaviour of the connectives when B is involved. So, for instance, $\neg\varphi$ will be simply true if and only if φ is simply false, and simply false if and only if φ is simply true. But what if φ is both true and false? We have already seen that the dialetheist accepts that a sentence is true if and only if its negation is false. Hence if φ is both true and false and receives the value B, $\neg\varphi$ will be false and true and hence also receive the value B. Similar considerations can be offered for conjunction, disjunction and the universal quantifier. As anticipated, the conditional is defined in terms of negation and disjunction, and the biconditional is defined in terms of conjunction and the conditional; finally, the existential quantifier is defined in terms of negation and the universal quantifier in the usual way. The upshot is a three-valued logic in which the semantics for connectives and quantifiers is the same as Kleene's (1952: §64) logic K_3 except that the middle value is designated.

In more detail and more formally, let $\mathfrak{A} = \langle \mathcal{D}, \mathcal{I} \rangle$, where $\mathcal{D}$ is a domain of objects and $\mathcal{I}$ an interpretation function. Given a variable assignment S (i.e. a mapping from variables to elements of $\mathcal{D}$), we can define the valuation function $v_S^{\mathfrak{A}}$ from formulae to truth-values as follows. (We omit the subscript and superscript when clear from context.) If φ is of the form $R_n(x_1, \ldots, x_n)$ then $v_S^{\mathfrak{A}}$ assigns to φ the value $\mathcal{I}$ assigns to $\langle R_n \langle Sx_1, \ldots, Sx_n \rangle \rangle$. If φ is of the form $\neg\psi$, $\psi \wedge \chi$ or $\psi \vee \chi$, then $v_S^{\mathfrak{A}}$ assigns a value to φ in conformity with the following matrices:

$\neg$			$\wedge$	T	B	F		$\vee$	T	B	F
T	F		T	T	B	F		T	T	T	T
B	B		B	B	B	F		B	T	B	B
F	T		F	F	F	F		F	T	B	F

Finally, if φ is of the form $\forall x \psi(x)$: (i) $v_S^{\mathfrak{A}}$ assigns the value T to φ just in case for every object d in D, $v_S^{\mathfrak{A}}$ assigns the value T to $\psi(d)$; (ii) $v_S^{\mathfrak{A}}$ assigns the value F to φ just in case for some object d in D, $v_S^{\mathfrak{A}}$ assigns the value F to $\psi(d)$; (iii) otherwise, $v_S^{\mathfrak{A}}$ assigns the value B to φ.

To conclude the formal characterization of LP we need to define logical consequence. As expected, we define it in terms of preservation of designated value. More formally, given an interpretation $\mathfrak{A}$ and a variable assignment S, we say that a valuation $v_S^{\mathfrak{A}}$ satisfies a sentence φ just in case $v_S^{\mathfrak{A}}(\varphi) \in \delta$, and that $v_S^{\mathfrak{A}}$ satisfies a set of sentences Σ just in case $v_S^{\mathfrak{A}}$ satisfies every element of Σ. Then we define semantic consequence thus:

$$\Sigma \models \varphi \text{ iff for every } \mathfrak{A} \text{ and } S \text{ whenever } v_S^{\mathfrak{A}} \text{ satisfies } \Sigma, v_S^{\mathfrak{A}} \text{ satisfies } \varphi.$$

LP's logical truths are the same as classical logic, but its consequence relation is significantly different. For instance, and quite importantly for our purposes, *ex falso quodlibet* fails: let $v(\varphi) = B$ and $v(\psi) = F$; then v satisfies $\{\varphi, \neg\varphi\}$ but not ψ. Using the same valuation, we can also show that, as required, disjunctive syllogism is not generally valid in LP, since v satisfies $\{\varphi, \neg\varphi \vee \psi\}$ but not ψ.

This concludes the exposition of the first-order Logic of Paradox LP. In the context of classical first-order logic, it is customary to single out a special *identity* predicate. We can do the same in the present context. In particular, we take $=$ to be a two-place predicate such that $x = y$ receives either the value T or the value B just in case $x = y$. The result is first-order Logic of Paradox with identity, or LP$_=$ for short. If, within an interpretation $\mathfrak{A}$, $x = y$ always receives either the value T or the value F, we say that identity is *standard* in $\mathfrak{A}$, but, to stress, LP$_=$ admits interpretations in which identity is *not* standard, but *glutty* – that is, interpretations in which $x = y$ receives the value B for some assignment to the variables. Analogously to the LP case, LP$_=$'s logical truths are the same as classical logic with identity, but its consequence relation is significantly different, and the counterexamples to *ex falso quodlibet* and disjunctive syllogism carry over to LP$_=$.

Naïve Set Theory in LP

Our next task is to examine what can and cannot be proved in NLP. The first systematic study of this is due to Restall (1992), who defines $x = y$ as $\forall z(x \in z \leftrightarrow y \in z)$. As expected, the theory is inconsistent, since it proves that there is at least one object – the Russell set – which both belongs and does not belong to itself (see Restall 1992: 424). But Restall's key result is that all the ZF axioms except Foundation are theorems of NLP (see Restall 1992: 427). This also means that the theory is not trivial: there is at least one sentence, *viz.* the Axiom of Foundation, which the theory does not prove.

However, as noted above, the consequence relation of LP is significantly different from that of classical logic. In particular, the fact that to avoid Curry's Paradox LP does not validate *modus ponens* means that many natural arguments used in standard set theory fail. To make things worse, besides *modus ponens*, transitivity of the material conditional and transitivity of the material biconditional also fail in LP. To see this, consider a valuation v such that $v(\varphi) = T$, $v(\psi) = B$ and $v(\chi) = F$. It is easy to check that (i) v satisfies $\{\varphi \to \psi, \psi \to \chi\}$ but not $\varphi \to \chi$; and (ii) v satisfies $\{\varphi \leftrightarrow \psi, \psi \leftrightarrow \chi\}$ but not $\varphi \leftrightarrow \chi$. This means that more natural arguments typically used in set theory fail. As a result, standard proofs of some central results of contemporary set theory break down. A case in point is Cantor's Theorem. For let $|a|$ be the cardinality of a. Then, the usual strategy to prove that

$|\mathcal{P}(a)|$ is greater than $|a|$ consists in supposing, for *reductio*, that there is a function f from a onto $\mathcal{P}(a)$. This means that we have

$$b = \{x \in a : x \notin f(x)\} = f(c)$$

for some $c \in a$. But then

$$c \in b \leftrightarrow c \notin f(c) \leftrightarrow c \notin b,$$

which, *by transitivity of the material biconditional*, implies a contradiction. But, as we have seen, transitivity of the material biconditional fails in LP.

To be sure, all of this does not rule out that alternative arguments for results whose standard proof does not go through in LP could be given. But in the absence of reasons for thinking that the proofs which seem to fail in NLP can be reconstructed in some other form, the prospects for developing a substantial amount of set theory in NLP look bleak, as Priest himself acknowledges (2006b: 250).

What is more, some of what NLP *can* prove depends on the way it treats identity, which is not uncontroversial. For recall that we are considering the case in which $x = y$ is defined as $\forall z(x \in z \leftrightarrow y \in z)$. Using this definition, Restall (1992: 427) shows that NLP proves that there are sets x and y such that $x \neq y$. The proof is simple: in each model there is at least one set – the Russell set – which both belongs and does not belong to itself; hence, by the definition of identity, there is a set which, besides being identical to itself, is also non-identical to itself. *A fortiori*, there is a set x which is non-identical to some set y.

This suggests that the existence of at least two objects might depend on the fact that we have defined identity so that it is glutty in *all* models. And in fact, Weir (2004: 393–398) shows that once identity is not defined *à la* Restall but taken as a semantic primitive (as in Priest 2006b), the existence of two objects no longer follows. In particular, there are models of $NLP_=$ in which identity is standard but $\exists x \exists y(x \neq y)$ only receives the value F. One might reply that the models where identity never receives the value B are pathological, and do not capture the full structure of the naïve universe of sets. But this is not very convincing: one would like to be given some principled reason for regarding these models as pathological. As Weir puts it:

Even if it is acceptable to admit models in which identity can be "glutty", we need a principled non-ad-hoc justification for excluding models in which identity is standard. How can it be wrong to interpret $x = y$ as true (only) when both terms refer to the same object and false (only) when they do not, especially since, in the models in question, it is definitely true in the metatheory that the referents are the same or definitely true that they are distinct? It will be highly unsatisfactory if the sole reason is to ensure that the proposition that there are at least two objects is entailed. After all, if I restrict admissible models to infinite ones then I can show that on my logic anything, $\exists x = x$, for example, 'entails' infinity. (Weir 2004: 397)

But let us set aside the issue of the models of naïve set theory for the moment, since we will return to it when discussing the model-theoretic strategy. Note that although it enables us to prove that there are at least two things, regarding identity as always glutty also comes at a price: the Indiscernibility of Identicals, $\varphi(x), x = y \vdash \varphi(y)$ fails. This, to stress, is the direction of Leibniz's Law which almost everyone accepts – where Leibniz's Law states that two objects are identical if and only if they have the same properties.

To see that the Indiscernibility of Identicals fails if we regard identity as always glutty, let $\varphi(x)$ be $x \in z$ and consider a S which assigns a to x and b to y; and let $v(x \in z) = B$ for every z-variant of S and $v(y \in z) = F$ when c is assigned to z; in fact, let S assign c to z. Then $\forall z(x \in z \leftrightarrow y \in z)$ receives the value B under the assignment S, since $v(x \in z) = B$ for every z-variant of it. The reason is that in LP a biconditional one of whose sides is both true and false is always true and false. But whilst $x \in z$ receives the value B under S, $y \in z$ receives the value F, which means that we have designated premises and an undesignated conclusion (see Weir 2004: 397–398).

Dialetheists might apply spin and reply that this, contrary to appearances, is not an unwelcome consequence of their theory. They might insist, for instance, that when dealing with the realm of the inconsistent, we should not expect the standard laws of identity to hold, and they could make their case stronger by pointing to other cases where it is appropriate to reject the Indiscernibility of Identicals. This seems doubtful, but, most importantly, does not get to the heart of the matter. Dialetheists claim that we should stick to the Naïve Comprehension Schema because of its intuitiveness. But to do this, we are asked to reject another principle which is hardly less intuitive than the Comprehension Schema itself. As a result, the initial appeal of a paraconsistent set theory based on the naïve conception seems seriously diminished.

Naïve Set Theory in Minimally Inconsistent LP

The following simple diagnosis of the proof-theoretic weakness of NLP$_=$ is tempting: by allowing gluts in an indiscriminate manner, the background logic of this theory allows for too many models. This diagnosis suggests the following course of action: restrict the range of admissible interpretations by moving to logics that are more restrictive in their tolerance of gluts. One such logic, due to Priest (2001), is *minimally inconsistent* LP (LPm for short). This is a non-monotonic logic whose notion of consequence is defined in terms of minimally inconsistent interpretations – interpretations which, roughly, are as consistent as possible. The idea is to order interpretations of LP in terms of how inconsistent they are and restrict LP to interpretations that are minimal with respect to this ordering. In a slogan: *tolerate gluts, but only when you absolutely have to*. The details do not really matter here, as long as we bear in mind two things. First, if a theory based on LP is non-trivial, then so is the same theory when the background logic is LPm,[4] which means that NLPm is non-trivial.[5] Second, any $LP_=$ consequence is also an LPm consequence, but there are some LPm consequences that are not $LP_=$ consequences. In particular, every classical consequence of a consistent set of premises is an LPm consequence, since in consistent situations the minimally inconsistent models are just the classical models.

Thus, whilst still invalidating *modus ponens* – as any logic fit for the material strategy ought to – LPm's consequence relation is more generous than LP's. For this reason, despite his pessimism about the prospects for the material strategy in $LP_=$, Priest (2006b) expressed optimism about the possibility of developing naïve set theory using LPm as a background logic.

However, it turns out that LPm's consequence relation is *too* generous. For say a set theory is *almost trivial* if it implies that $\forall x \forall y (x \in y \wedge x \notin y)$. Thomas (2014: §4) has shown that NLPm's only model is one in which $\forall x \forall y (x \in y \wedge x \notin y)$ holds and identity is standard. Thus, if it has models, NLPm is almost trivial. Almost triviality is not the same as triviality but comes very close to it: triviality is usually considered unacceptable because in its presence no discrimination is possible with regards to what follows from what; almost triviality seems unacceptable because in its presence no discrimination is possible with regards to what belongs to what. Moreover, NLPm has only one model up to isomorphism, namely a one-element model. This means that whilst proof-theoretically too generous, NLPm is model-theoretically too restrictive: to take the universe of sets to consist of one, inconsistent set is to relinquish any interesting use of set theory, not only as a foundation for mathematics.

Can alternative versions of LPm fare any better as a background logic for naïve set theory on the material strategy? This is an open question, but note that all extant versions of the logic give rise to theories that, like NLPm, are either too weak or too

---

[4] To be precise, there are certain restrictions on this latter claim, but they are irrelevant for present purposes, so I will ignore them. The interested reader can consult Priest 2006b: 226–228.

[5] Spelling this out a bit more, it is easy to see that Restall's non-triviality proof carries over to $NLP_=$, which then implies that NLPm is non-trivial.

strong, as Thomas (2014) has also shown. In particular, three alternative versions of LPm have been developed by Marcel Crabbé (2001). Using one of them as a background logic, the situation is exactly the same as for NLPm: the theory has one model up to isomorphism, consisting of only one element, and the theory is almost trivial. Using the other two alternative versions of LPm as background logics we avoid almost triviality but at the price of not improving upon $LP_=$ as a background logic: the resulting naïve set theories cannot prove the existence of sets that behave like singleton sets, sets that behave like ordered pairs, and sets that behave like infinitely ascending linear orders.

For the reasons pointed out in this section, I conclude that the prospects for developing a naïve set theory by pursuing the material strategy look rather dim, and those for developing it on the basis of LP and cognate systems even dimmer. We need to turn to the relevant strategy, and see whether it fares any better.

4.4 The Relevant Strategy

The relevant strategy, to repeat, takes the conditionals which appear in the Axioms of Comprehension and Extensionality to be not material conditionals, but rather conditionals defined with the help of some kind of relevant logic validating *modus ponens*. The biconditional is then defined in the usual way.[6]

But what kind of motivation could we have for this strategy? Priest seems to have in mind something like the following. In the previous section, we saw that the dialetheist thinks that the classical logician errs in assuming that there are no sentences that are both true and false. We then showed that, once this assumption is rejected, the truth-tables for negation, conjunction and disjunction in LP can be motivated quite naturally on the basis of elementary principles relating truth to falsity and vice versa. We then went on to define the conditional in terms of negation and disjunction. But, Priest seems to think, in so doing we have followed the classical logician in *another* of her mistakes. Not only do classical logicians err in assuming that no sentence can be both true and false; they also err in assuming that the conditional has to be characterized as the material conditional: the conditional has to satisfy some relevance constraints. The upshot seems to be that our logic will have to be like LP except that we do not define $\varphi \to \psi$ as $\neg \varphi \vee \psi$, but, rather, give a relevant account of the conditional.

---

[6] Although, as we shall see below, the biconditional which appears in the Axiom of Extensionality might have to be defined in a different way.

Standardly, a propositional logic is *relevant* just in case whenever $\varphi \to \psi$ is logically valid, φ and ψ have a propositional parameter in common: the guiding idea is that for a conditional to be true, there needs to be some content shared between the premiss and the conclusion. The obvious initial problem the relevant strategy has to face is that the strengthened version of Curry's Paradox shows that it is not enough for the conditional to be relevant in the usual sense to avoid triviality: the standard relevant logic R (Anderson and Belnap 1975), for instance, satisfies Contraction (besides *modus ponens*), and is therefore too strong for the purposes at hand.

This means that if the relevant strategy is to have any mileage, one must consider logics which do not validate Contraction and the Modus Ponens Axiom but retain *modus ponens*. The logic that Priest (2006b: 251) asks us to consider to this end is one of the so-called *depth relevant* logics. The name is due to the fact that these logics can be shown to satisfy the so-called *depth relevance condition*, introduced by Brady (1984). To explain this condition, we need to explain the notion of *depth*, which can be recursively defined as follows:

(1) The subformula φ of the formula φ is of depth 0.
(2) If $\neg\psi$ is a subformula occurrence of φ of depth d in φ, then this occurrence of ψ is of depth d in φ.
(3) If $\psi \wedge \chi$ is a subformula occurrence of φ of depth d in φ, then these occurrences of ψ and of χ are both of depth d in φ.
(4) If $\psi \to \chi$ is a subformula occurrence of φ of depth d in φ, then these occurrences of ψ and of χ are both of depth $d + 1$ in φ.

Thus, the depth of an occurrence of a subformula in a formula φ measures the number of nestings of conditionals under which the occurrence of the subformula appears. A logic satisfies the depth relevance condition just in case whenever $\varphi \to \psi$ is logically valid, φ and ψ have a propositional parameter of the same depth in common. The idea is that although the axiom form of Contraction ($\varphi \to (\varphi \to \psi)) \to (\varphi \to \psi)$ cannot be faulted on grounds of irrelevance – φ and ψ occur in both the antecedent and the consequent – the axiom should be faulted because ψ has a different depth in the antecedent and in the consequent: in the antecedent it has depth 2 (since it is under two conditionals), whereas in the consequent it has depth 1. It is easy to see why from this perspective the Modus Ponens Axiom is not valid either: whilst ψ has depth 1 in the antecedent, it has depth 0 in the consequent.

As partly anticipated, the particular logic considered by Priest is the result of putting together LP's account of the extensional connectives with a depth relevant account of the conditional based on a modal semantics, which we now need to sketch (for details, see Priest 2006b: 270–271). We have two disjoint sets of worlds, the logically possible and the logically impossible. Just as there are physically impossible worlds – worlds where the laws of physics are different – there are (logically) impossible worlds – worlds where the laws of logic are different. The actual world belongs to the possible worlds, and validity is defined as preservation of truth at just the possible worlds. Moreover, it is stipulated that there are no truth-value gaps in the possible worlds, which ensures that the logic validates the Law of Excluded Middle, $\varphi \vee \neg\varphi$.

The truth and falsity conditions for the extensional connectives are as expected: they are the same as in LP, except that they are relativized to the appropriate world. The conditional which concerns us, on the other hand, is characterized in two steps.

First, we define $\rightarrow$ in terms of a non-contraposing conditional $\Rightarrow$. Thus, $\varphi \rightarrow \psi$ is $\varphi \Rightarrow \psi \wedge \neg\psi \Rightarrow \neg\varphi$. This ensures that Contraposition, $\varphi \rightarrow \psi \vdash \neg\psi \rightarrow \neg\varphi$, holds for the conditional which is going to figure in the Axioms of Comprehension and Extensionality.

Then, we set things up so that the truth of $\varphi \Rightarrow \psi$ at a possible world requires preservation of truth at all worlds, including the impossible ones. This means, for instance, that $\varphi \Rightarrow \varphi$ is true at the actual world. The falsity of conditionals which are not depth relevant is then due to the fact that conditionals may behave in very odd ways at impossible worlds. More specifically, the truth of $\varphi \Rightarrow \psi$ at an impossible world w requires preservation of truth *across* worlds standing in a certain relation R to w: $\varphi \Rightarrow \psi$ is true at w if and only if for all worlds x, y such that $Rwxy$, if φ is true at x then ψ is true at y.

Thus, for instance, $\varphi \Rightarrow (\psi \Rightarrow \psi)$ will be false at the actual world because $\psi \Rightarrow \psi$ may fail at impossible worlds. (Just take an impossible world w standing in the relation R to a world x where ψ is true and to a world y where it is false.) The failure of Contraction at the actual world can be explained along similar lines: $\varphi \Rightarrow (\varphi \Rightarrow \psi)$ can be true at the actual world and $\varphi \Rightarrow \psi$ false because there can be impossible worlds where *modus ponens* fails, so that φ and $\varphi \Rightarrow \psi$ hold at them but ψ does not.

We have mentioned earlier that Priest's motivation for adopting the relevant strategy seems to be that he thinks that the conditional should satisfy some relevance constraints. But what reasons could we have for taking the conditional to be depth relevant? Brady (2003; 2006), the originator of depth relevant logics, offers considerations based on *meaning containment*. His starting point is that a conditional such as $\varphi \rightarrow \psi$ is logically true just in case the content of ψ is contained in that of φ. However, before one can even begin to evaluate which particular logic is pinned down by such a proposal, an account of what a content is, and of what it is for a content to be contained in another, seems to be required. And this Brady does not provide.

Moreover, one would like to be told *why* we should regard an entailment as true just in case the content of the antecedent is contained in that of the consequent. All Brady (2003: 272) says is that we should regard conditionals as 'meaning containment statements', but some argument seems to be needed for doing so, especially given that it would seem to involve a radical departure from any prior conception of a conditional we might have.

Finally, even if we accept Brady's arguments, it seems clear that they do *not* pin down the logic considered (and needed) by Priest. Consider, for instance, the *rule* form of Contraction (Contr). This is to be rejected on the relevant strategy, if we are to avoid triviality. But what argument could Brady give for that? His general strategy is to say that 'meaning containments should not apply between containments that are not alike or between containments and non-containments' (2003: 272). Now is this argument to be extended to inferences or not? If it is, then the rule form of Contraction does seem to fail, by extending the argument Brady offers for the axiom form of Contraction, since in moving from $\varphi \Rightarrow (\varphi \Rightarrow \psi)$ to $\varphi \Rightarrow \psi$ we are moving from a containment between a general content and a containment to a containment, to use Brady's terminology. But then, for similar reasons, we should not be able to move from $\varphi \rightarrow \neg\varphi$ to $\neg\varphi$ either, since we would be moving from a containment to a non-containment. But the inference in question is validated by Priest's semantics. If, on the other hand, Brady's argument is not to be extended to inferences, then it is not clear why our logic should not validate Contraction. It will not do to say that the rule form of Contraction should fail because the corresponding axiom form does. For, by parity of reasoning, the axiom form of the rule $\varphi \rightarrow \neg\varphi \vdash \neg\varphi$ should fail too, since the corresponding axiom form fails according to Brady's argument. Neither will it do to say that the rule form of Contraction, and not the rule $\varphi \rightarrow \neg\varphi \vdash \neg\varphi$, is logically equivalent to the corresponding axiom form. For this assumes that logical equivalence under the logic in question is correct, which is precisely part of what we are trying to motivate.

Priest, however, makes no mention of Brady's motivation. Instead, he seems to oscillate between two different motivations. The first motivation is that a depth relevant conditional solves the paradoxes of material implication. As pointed out by Field (2008: 374), however, this is not a convincing motivation for a depth relevant conditional. Standard examples of oddities to which the material conditional supposedly gives rise are those of conditional statements which have a false antecedent or a true consequent, and are therefore true if the conditional is material.

The conditional of the logic Priest is considering does deal with this sort of cases, although so seemingly do a number of other proposals which have been made. We do not need to review these other proposals here, however. For the crucial point is that depth relevant conditionals obey strengthening of the antecedent, so that we can infer from $\varphi \to \psi$ to $(\varphi \wedge \chi) \to \psi$. So, for instance, the relevant conditional Priest is considering allows us to infer from 'If that's a dog, it has four legs' to 'If that's a dog that has a leg amputated, it has four legs'. And this sort of inference is generally regarded to be just as problematic as the cases depth relevant conditionals are intended to deal with.

The second motivation Priest seems to have in mind for his conditional is closely related to its modal semantics. He suggests that the conditional $\to$ is the connective of entailment, so that its intended reading is 'if ..., then logically ...' or 'if ..., then it follows logically that ...' (see Priest 2006b: 86–88). Call this the *logical* reading of the conditional. He then goes on to argue that the characterization of $\to$ that he offers can be motivated on the basis of this reading. Two features of it are especially salient. The first feature is that a logical conditional should preserve truth forwards and falsity backwards: a true entailment, Priest claims, preserves truth from antecedent to consequent and falsity from consequent to antecedent. This is the reason why we should characterize $\to$ in terms of $\Rightarrow$ so that it is contraposible.[7] The second feature is that a true entailment should preserve truth and falsity in the appropriate directions *necessarily*, that is at all worlds, both the possible and the impossible ones. This is why the truth-conditions for $\to$ need to be given in terms of the modal semantics sketched earlier.

Does the second motivation for taking the Axioms of Comprehension and Extensionality to be depth relevant fare any better? Notice, for a start, that even if Priest's arguments for taking the connective of entailment to be characterized by the conditional $\to$ are successful, they still do not address the prior question of why we should take the conditional in the Axioms of Comprehension and Extensionality to be the connective of entailment. This means that Priest still owes us an argument

---

[7] In at least one place (2006b: 270), Priest claims that it is the conditional $\Rightarrow$ which is to be read as 'if ..., then logically ...'. This raises the question of what to make of the argument that a true entailment, besides truth preservation from antecedent to consequent also requires falsity preservation from consequent to antecedent. The reason for Priest's oscillation here is due, I suspect, to the fact that he needs the conditional in the Axioms of Comprehension and Extensionality to be contraposible – otherwise the resulting set theory is too weak – but wants the one he uses in his theory of truth to be non-contraposible, for reasons we cannot go into here – see Priest 2006b: 69–72. An alternative strategy he might adopt would consist in rejecting his argument that the logical conditional is non-contraposible and claiming that the versions of the Axioms of Comprehension and Extensionality obtained by defining $\to$ in terms of $\Rightarrow$ are still intuitive. Since Priest's theory of truth is not among our concerns here, however, we can set this problem aside and take Priest as claiming that it is the conditional $\to$ which is to be read as the logical conditional.

to the effect that the Axioms of Comprehension and Extensionality which his logic validates are those which, he claims, we find intuitive. In other words, it is one thing to claim that, besides thinking that no sentence can be both true and false, the classical logician errs in taking the conditional to be the material implication because this cannot possibly be our conditional, in that it gives rise to the paradoxes of material implication. It is another thing to claim that we should take the conditional to be an entailment without providing any argument that entailment bears some resemblance to the conditional we employ when motivating the axioms of set theory. Until such an argument is provided, it is unclear that what we are validating really are the principles which we are trying to validate.

But not only does the logical reading of the conditional make it unclear whether the principles the dialetheist is trying to validate really are the principles which, according to her, we find intuitive; it also makes it mysterious on which grounds she can claim that she *is* validating them.

Consider, for instance, the Axiom of Comprehension. Given the definitions of $\leftrightarrow$ and $\rightarrow$, this is true only if

$$\exists y \forall x (\varphi(x) \Rightarrow x \in y) \tag{4.1}$$

is also true.[8] (4.1), in turn, is, in Priest's logic, true just in case for some a belonging to the domain of interpretation, $\forall x (\varphi(x) \Rightarrow x \in a)$. But, given the truth-conditions for $\Rightarrow$ at possible worlds, it is easy to see that the truth of this latter claim requires that at all worlds w – including the impossible ones – we have that if $\varphi(b)$ is true then so is $b \in a$ (where b is an object in the domain of interpretation).

But why believe that? Priest does not address this sort of worry, but it seems that he should. For his explanation of the failure of Contraction – needed to avoid triviality – depends on allowing impossible worlds into our interpretation. Setting aside the technical definition of the conditional at impossible worlds, the problem, more generally, is that we are asked to suppose that the conditional, whilst behaving at impossible worlds in such a way that $\varphi \rightarrow \varphi$ may fail at them, also behaves in such a way that some substantial mathematical facts involving it are maintained. The notion of an impossible world seems to have become utterly mysterious, so as to jeopardize its role in explaining the failure of Contraction.[9]

---

[8] For we have that $\exists y \forall x (x \in y \leftrightarrow \varphi(x))$ is true at the actual world iff $\exists y \forall x ((x \in y \rightarrow \varphi(x)) \wedge (\varphi(x) \rightarrow x \in y))$ is true at the actual world iff $\exists y \forall x (((x \in y \Rightarrow \varphi(x)) \wedge (\neg \varphi(x) \Rightarrow x \notin y) \wedge ((\varphi(x) \Rightarrow x \in y) \wedge (x \notin y \Rightarrow \neg \varphi(x)))))$ is true at the actual world.

[9] It is perhaps worth noticing that this problem does not automatically affect Priest's use of a logical conditional in other areas. For instance, in the case of his theory of truth it seems more plausible (especially on certain conceptions of truth, such as deflationism) to suppose that, for all possible worlds, including the impossible ones, if φ is true at them, then $T(\varphi)$ is also true at them, and vice versa (where T is the truth-predicate). For an account of truth which is both dialetheist and deflationist, see Beall 2009.

Thus, Priest does not seem to have offered a good argument for focusing on the logic he is considering, which would be enough to cast serious doubts on the viability of the relevant strategy. But there is a further point which deserves mention, namely that *even if* we grant that the conditional should be depth relevant, we still face the problem that a conditional of this kind is *very* weak. We can define the conditional which appears in the Axioms of Comprehension and Extensionality in terms of a more primitive conditional $\Rightarrow$, so that it satisfies Contraposition; still, the resulting system does not suffice to carry out many of the standard arguments of contemporary set theory, especially those employing *reductio*. A case in point is, again, Cantor's Theorem (see Priest 2006b: 253). Suppose φ is an assumption made for the purposes of *reductio*, and ψ is the conjunction of other facts appealed to in the *reductio*. Even if we establish that $(\varphi \wedge \psi) \rightarrow (\chi \wedge \neg\chi)$, by contraposing and detaching we can only get $\neg\varphi \vee \neg\psi$. But even given ψ, we cannot get $\neg\varphi$ because we do not have disjunctive syllogism.

To overcome this sort of problems, Weber (2010b; 2012; 2013) has suggested further strengthening the conditional, compatibly with the restrictions required to avoid triviality. In particular, he suggests extending Brady's dialectical relevant logic DKQ – a kind of depth relevant logic – with the Counterexample Rule $\varphi, \neg\psi \vdash \neg(\varphi \rightarrow \psi)$. The resulting logic DLQ gives rise to the naïve set theory NDLQ – here, the Naïve Comprehension Schema also encompasses instances in which y occurs free in the defining predicate φ. Brady (1989) showed that naïve set theory based on DKQ is inconsistent but non-trivial, and his construction also covers NDLQ.[10]

It is not clear that NDLQ has the resources required to carry out most of the constructions of standard set theory, but the prospects certainly look brighter than those for NLP and cognate systems. In particular, Weber (2010b; 2012) shows how to reconstruct several basic results about cardinal and ordinal numbers within NDLQ. He defines ordinals as well-ordered irreflexive transitive sets connected by $\subseteq$. Given this definition one can prove a principle of transfinite recursion and the existence of the set of all ordinals. This set can be well-ordered, and this induces a well-ordering on the set of all sets, from which the Axiom of Choice follows. Given Choice, one can then define cardinals using the von Neumann cardinal assignment. On the basis of this definition, it can be shown that NDLQ proves the existence

---

10 To be precise, Weber (2012) uses the Counterexample Rule, whilst Weber (2010b; 2013) uses the Counterexample Axiom $(\varphi \rightarrow \psi) \rightarrow \neg(\varphi \wedge \neg\psi)$, whose contraposed form is $(\varphi \wedge \neg\psi) \rightarrow \neg(\varphi \rightarrow \psi)$. The non-triviality of naïve set theory in DKQ together with the Counterexample Axiom does not follow from Brady's construction. On the other hand, the results in Weber 2010b still follow using the Counterexample Rule, *modulo* adding a condition in the definition of an ordinal which can simply be proved using the Counterexample Axiom.

of various large cardinals (such as inaccessibles, Mahlos and even measurables[11]) as well as Cantor's Theorem. The particular argument given for Cantor's Theorem makes essential use of the Counterexample Rule, which makes vivid the utility of its addition to the logic.

But is there any independent motivation for the new logic? Until such a motivation is provided, its adoption looks ad hoc. Obviously, the Counterexample Rule appears prima facie plausible: whenever the antecedent of a conditional is true and its consequent false, the conditional is false. But then, disjunctive syllogism also appears prima facie plausible, but almost no paraconsistent logic can include it, on pain of triviality. The prima facie plausibility of being able to infer ψ from φ and $\neg\varphi \vee \psi$ is usually explained by saying that one does not immediately consider the case in which φ is a dialetheia. Similarly, however, the Counterexample Rule seems much less plausible when dialetheias enter the picture: whenever φ is a dialetheia, the rule allows one to infer that something as seemingly obvious as $\varphi \rightarrow \varphi$ is false.[12]

Now Brady and Priest offered arguments for the conditional of depth relevant logic. However, these arguments do not carry over to DLQ. For a moment's reflection shows that the Counterexample Axiom is not depth-relevant: φ and ψ have both depth 2 in the antecedent and depth 1 in the consequent. This does not tell immediately against the acceptability of its rule form from a depth relevantist point of view, but it tells us that such an argument cannot come from the depth relevance condition. As for Priest's idea that $\rightarrow$ is the connective of entailment, this seems problematic once the conditional obeys the Counterexample Rule in the presence of dialetheias. For, as already observed, when φ is a dialetheia, the Counterexample Rule allows us to conclude $\neg(\varphi \rightarrow \varphi)$. This means that φ is not a logical consequence of φ. It is hard to see why this is any less unacceptable than rejecting the Naïve Comprehension Schema.

Weber, however, argues that we can justify the adoption of DLQ as the background logic for naïve set theory on the basis of the fact that it allows us to recapture

---

[11] A cardinal κ is *measurable* iff $\kappa > \omega$ and there is *measure* on κ, i.e. there is a function $\mu : \mathcal{P}(\kappa) \mapsto \{0, 1\}$ such that:
(i) if $\{\chi_i : i \in \iota\}$ is a pairwise disjoint family of subsets of κ and the cardinality of ι is less than κ, then

$$\bigcup_{i \in \iota} \chi_i = \sum_{i \in \iota} \mu(\chi_i);$$

(ii) $\mu(\kappa) = 1$ and $\mu(\{\alpha\}) = 0$ for $\alpha < \kappa$.
Intuitively, μ is to be thought of as selecting out the 'big' subsets of κ, where a set of μ-measure 0 is 'small' and a set of μ-measure 1 is 'big'. Clause (i) guarantees that the union of fewer than κ 'small' subsets of κ is still 'small'. Clause (ii) guarantees that κ is 'big' and that, for any α less than κ, the singleton $\{\alpha\}$ is 'small'.

[12] Indeed, as Brady (2014: 280) observes, the standard justification of the Counterexample Rule makes use of classical considerations. In contraposed form, *modus ponens* says that if $\vdash \varphi$ but $\nvdash \psi$ we have that $\nvdash \varphi \rightarrow \psi$. Now suppose $\vdash \varphi$ and $\vdash \neg\psi$. If ψ is not a dialetheia, from $\vdash \neg\psi$ it follows that $\nvdash \psi$. This, together with $\vdash \varphi$, delivers, by the contraposed form of *modus ponens*, that $\nvdash \varphi \rightarrow \psi$. If $\varphi \rightarrow \psi$ is negation-complete, it follows that $\neg(\varphi \rightarrow \psi)$, as desired.

central results of Cantorian set theory. His approach is summed up in the following remark:

Essentially, we want the strongest logic possible that does not explode when given a comprehension principle. (Weber 2010b: 73)

But the problem is that Weber is not entitled to talk of *the* strongest logic possible that does not explode in the presence of the Comprehension Schema, at least if strength is understood in terms of what the theory can prove (which seems to be the sense used throughout his writings on the subject). To take a simple example, Field et al. (2017) generalize and extend Brady's abovementioned construction to show the non-triviality of several naïve set theories. Some of these set theories avoid triviality by placing restrictions on negation or going dialetheic. Others avoid triviality by invalidating the Weakening Rule $\psi \vdash \varphi \rightarrow \psi$. Neither of these two types of background logic is stronger than the other, but admitting principles from both of them in naïve set theory leads to triviality.[13] Which one is the correct naïve set theory? What we end up with is a number of different logics, each of which enables us to prove certain results, and for none of which we seem to have a good motivation: any attractions of the paraconsistent solution to the set-theoretic paradoxes seem seriously undermined.

The point is exacerbated by the way Priest (see 2006b: 254) deals with another problem the relevant strategy has to face. The problem is that the strategy runs the risk of undermining the naïve theory's credentials for being a theory of *sets*. For if φ is a true sentence, then the left-to-right direction of $x \in y \leftrightarrow (x \in y \wedge \varphi)$ is not generally relevantly valid. This means that even though y and $\{x : x \in y \wedge \varphi\}$ have the same members, we do not have $y = \{x : x \in y \wedge \varphi\}$. Ultimately, this is due to the fact that the biconditional which appears in the Axiom of Extensionality is now being treated as an intensional operator. A consequence of this, however, is that although the Axiom of Extensionality holds in the theory, that does not guarantee that sets are extensional entities, which is arguably part of the set concept (see Section 1.4).

To overcome this problem,[14] Priest suggests replacing the biconditional in the Extensionality Axiom with a biconditional corresponding to an enthymematic

---

[13] In the next chapter, we will see that a structurally similar problem affects the attempt to restrict the Naïve Comprehension Schema according to consistency maxims.

[14] Notice that it will not do to replace the biconditional in the Extensionality Axiom with the material biconditional. For, as already pointed out in the previous section, in LP a biconditional one of whose sides is a dialetheia is always true and false, and hence true. So let φ be a dialetheia. Then for any set a, we have that $\forall x(x \in a \leftrightarrow \varphi(x))$. By Extensionality, we can conclude that there is only one set – the set $a = \{x : \varphi(x)\}$. See Priest 2006b: 254, although I have slightly modified Priest's argument, which seems to postulate a membership relation holding between the members of a and the condition φ.

conditional $\rightarrow$.[15] This is defined in two steps. First, we add to our system a logical constant t, which is to be thought of as the conjunction of all truths, and which validates the following inferences:

$$\vdash t$$

$$\varphi \vdash t \rightarrow \varphi$$

Then we define $\varphi \rightarrow \psi$ as $(\varphi \wedge t) \rightarrow \psi$. Priest's proposal is then to formulate the Axiom of Extensionality as $\forall x (x \in y \rightleftharpoons x \in z) \rightarrow y = z$, where $\rightleftharpoons$ is the biconditional corresponding to $\rightarrow$. This, Priest claims, overcomes the problem raised above. For let φ be any truth. Then, since $t \rightarrow \varphi$, we have that $x \in y \rightarrow (x \in y \wedge \varphi)$. But since we obviously also have that $(x \in y \wedge \varphi) \rightarrow x \in y$, we can conclude that y and $\{x : x \in y \wedge \varphi\}$ are identical.

Setting aside the problems raised by the interpretation of t, all of this makes it even less clear whether the Axioms of Comprehension and Extensionality that we are validating on the relevant strategy are the ones which, according to the naïve set theorist, we find intuitive. For what we now have is an Axiom of Extensionality with two different conditionals, for neither of which we have an argument that they correspond to any conditional we had a prior conception of.

Worse still, Weber (2010a) has shown that, in the context of NDLQ, formulating Extensionality using an enthymematic conditional leads to triviality. Weber argues that the proponent of the relevant strategy should not have been troubled by this problem to begin with. For, he says, if φ is a dialetheia, then we *should not* have $y = \{x : x \in y \wedge \varphi\}$, since y and $\{x : x \in y \wedge \varphi\}$ will *not* have the same members: the objects that y only includes are both included and excluded by y and $\{x : x \in y \wedge \varphi\}$. However, by dialetheist lights, this should mean that y and $\{x : x \in y \wedge \varphi\}$ both have and do not have the same members, and so we *should* be able to conclude that $y = \{x : x \in y \wedge \varphi\}$. What is more, the logic does not allow us to conclude that $y = \{x : x \in y \wedge \varphi\}$ even when φ is not a dialetheia, which means that the theory cannot prove the identity of collections for which it is uncontroversially the case that they have the same members.

In sum, the relevant strategy, at least as developed by Weber, enables the paraconsistent set theorist to make some progress over the material strategy, since NDLQ has the resources to carry out some reasonable amount of set theory. However, the theory has a background logic which is poorly motivated. And, on pain of triviality,

---

[15] The enthymematic conditional used by Priest is similar to the one defined by Beall et al. (2006) to deal with restricted quantification in relevant logic.

the current attempts to provide it with a genuine principle of extensionality fail, so that it cannot be regarded as a *set* theory.

4.5 The Model-Theoretic Strategy

Perhaps for these reasons, however, Priest's favourite strategy for developing a naïve set theory seems to be what he calls the *model-theoretic strategy* (see Priest 2006b: §18.4, Priest 2017: §11). Let us first outline the strategy, and then look at it in more detail.

The guiding idea here is that, for a dialetheist, classical logic is fine when reasoning over consistent domains: it is only in the domain of the inconsistent that we should use different logics, since it is only there that contradictions arise. At least, this is the view of the kind of dialetheist whom we encountered in Section 4.3 and who makes use of the logic LP$_=$. According to dialetheists of this kind, the only mistake of the classical logician is to suppose that no sentence can be both true and false. Thus, when they reason in a context where no sentence is true and false, they may make use of classical logic. The analogy is with the case of the intuitionist logician, who can feel free to use classical logic when reasoning over decidable domains.

Now, when applied to the case of set theory, this suggests a different strategy for developing a naïve set theory which is not too weak: rather than just deriving things from the Axioms of Comprehension and Extensionality in some paraconsistent logic, the dialetheist can also restrict attention to a consistent subdomain of the naïve universe of sets, and use classical logic in deriving things from axioms describing that subdomain. But which subdomain? Priest's idea is that the cumulative hierarchy is such a subdomain, and he sets out to show how this could be by proving that there are models of ZF which are also models of NLP$_=$ and contain the cumulative hierarchy up to the first inaccessible cardinal. Thus, according to Priest, the naïve universe of sets models the axioms of ZF, which, therefore, the dialetheist can regard as true. Hence, she can use them to prove set-theoretic results, and, in so doing, she can use classical logic, since what they describe is a consistent domain.

This is the general flavour of the proposal. To examine it in more detail, we need to look at Priest's model-theoretic construction. So consider again NLP$_=$, the axiomatization where we have Extensionality and Comprehension and the conditional is the material conditional. We want to construct models of NLP$_=$ which are also models of ZF. To this end, we make use of a more general result.

In particular, let $\mathfrak{A}$ be $\langle \mathcal{D}, \mathcal{I} \rangle$ and v be the value assignment under $\mathcal{I}$. Moreover, let $\sim$ be any equivalence relation on $\mathcal{D}$, and, if $d \in \mathcal{D}$, let $[d]$ be the equivalence class induced by $\sim$ containing d. Then we can define a *collapsed model* $\mathfrak{A}^{\sim} = \langle \mathcal{D}^{\sim}, \mathcal{I}^{\sim} \rangle$ with assignment $v^{\sim}$ by letting $\mathcal{D}^{\sim}$ be $\{[d]: d \in \mathcal{D}\}$ and characterizing $\mathcal{I}^{\sim}$ as follows. First, if c is a constant which denotes d in $\mathfrak{A}$, it denotes $[d]$ in $\mathfrak{A}^{\sim}$. Second, if f is an n-place function symbol which denotes d in $\mathfrak{A}$ when taking as arguments terms denoting $[d_1], \ldots, [d_n]$, then it denotes $[d]$ in $\mathfrak{A}^{\sim}$ when taking as arguments terms denoting $d_1, \ldots, d_n$. Third, if P is an n-place predicate, we let $\langle a_1, \ldots, a_n \rangle$ be in P's positive (negative) extension in $\mathfrak{A}^{\sim}$ iff $\exists x_1 \in a_1, \ldots, \exists x_n \in a_n$ such that $\langle x_1, \ldots, x_n \rangle$ is in the positive (negative) extension of P in $\mathfrak{A}$. In other words, given a model, we can construct a collapsed model by inducing an equivalence relation on the members of its domain and identifying the member of $\mathcal{D}$ in the same equivalence class, so that, from $\mathfrak{A}^{\sim}$'s perspective, the equivalence classes are individuals having the properties of their members. Given this construction, it is then easy to prove the following result:[16]

Lemma 4.5.1 (Collapsing Lemma). *For every formula φ in the language of* $\mathfrak{A}$, $v^{\sim}(\varphi) \supseteq v(\varphi)$.[17]

The Collapsing Lemma guarantees that any model $\mathfrak{A}$ can be collapsed into a model $\mathfrak{A}^{\sim}$ in such a way that no formula loses a truth-value, but some formulae can gain one. In other words, formulae can go from being simply true or false to being *both* true and false. We say that the resulting model is collapsed because we restrict the original model by taking the objects to be the equivalence classes under the given equivalence relation.

The next step is to collapse a model $\mathfrak{A} = \langle \mathcal{D}, \mathcal{I} \rangle$ of ZF into a model $\mathfrak{A}^{\sim} = \langle \mathcal{D}^{\sim}, \mathcal{I}^{\sim} \rangle$ of ZF and NLP$_=$. What is crucial is the choice of $\sim$, and what we are after is a model which contains the cumulative hierarchy but validates the Comprehension Schema. The crucial observation is that, in LP$_=$, if in a biconditional $\varphi \leftrightarrow \psi$, φ is both true and false, then the biconditional is true (and false) no matter what the truth value of ψ is (as briefly remarked in Sections 4.3 and 4.4, fn. 14). This means that if we can construct a collapsed interpretation with an object $[a]$ designated by a such that $x \in a$ receives the value B for all x, then $\forall x (x \in a \leftrightarrow \varphi)$ will be true, and so will $\exists y \forall x (x \in y \leftrightarrow \varphi)$: our model will be a model of NLP$_=$ (since the original model was already a model of Extensionality). But this is straightforward: just define an equivalence relation in such a way that

---

[16] The proof is a straightforward induction on the complexity of formulae, and can be found in Priest 2006b: 229–230.

[17] It is perhaps worth recalling that in the framework under consideration, valuations are officially maps to *sets* of values. This explains why the Lemma is formulated as saying that, for any φ in the language of $\mathfrak{A}$, $v(\varphi)$ is a *subset* of $v^{\sim}(\varphi)$.

all the objects occurring up to a level V_α keep their original interpretation and all other objects get identified. In particular we let α be any ordinal in $\mathfrak{A}$, a be V_α, and define $\sim$ as follows:

$(x$ and y are in a (in $\mathfrak{A})$ and $x = y)$ or $(x$ and y are not in a (in $\mathfrak{A}))$.

All we need to check is that by applying the Collapsing Lemma we obtain a model $\mathfrak{A}^\sim$ in which, for all $b \in \mathcal{D}^\sim$, $b \in a$ receives the value B and is, for the reasons given above, also a model of NLP$_=$. So suppose that b is of rank less than α. Then $b \in a$ is true in $\mathfrak{A}$ and so is in $\mathfrak{A}^\sim$. Conversely, suppose that b is not of rank less than α. Then, it belongs to an x which is also not of rank less than α and must therefore be in $[a]$, which means that $b \in a$. This means that $b \in a$ always receives the value T in $\mathfrak{A}^\sim$. But since whatever b is, there are elements of rank not less than α such that b is not an element of them and which have been identified with a in $\mathfrak{A}^\sim$, $b \in a$ also receives the value F in $\mathfrak{A}^\sim$.

So $\mathfrak{A}^\sim$ is a model of both ZF and NLP$_=$, and consists of a segment of the cumulative hierarchy up to α and a glutty object $[a]$ such that $x \in a$ receives the value B for all assignments to x. Our last step is to construct a collapsed model containing the cumulative hierarchy up to the first inaccessible cardinal. Again, this is straightforward. Just take a model of ZF in which there are at least two inaccessibles, κ_1 and κ_2, with $\kappa_2 > \kappa_1$. Then set α to be κ_2 and apply the construction described above. The resulting model $\mathfrak{A}^\sim$ is a model of ZF, is a model of NLP$_=$, and contains the cumulative hierarchy up to κ_1 as a consistent inner model.

But what does this show? Priest's model-theoretic strategy is to use this construction to motivate the paraconsistent acceptability of all the classical theorems of ZF. The fact that we can construct a model of NLP$_=$ which contains the cumulative hierarchy up to the first inaccessible as a consistent inner model is used to support the view that we can use classical logic when deriving things from ZF: the use of classical logic is, even from the paraconsistent viewpoint, acceptable when reasoning about consistent domains, such as, presumably, the cumulative hierarchy.

Priest concludes:

Since the universe of sets is a model of ZF (as well as naive set theory), these hold in it. We may therefore establish things in ZF in the standard classical way, knowing that they are perfectly acceptable from a paraconsistent perspective. We cannot, of course, require the theorems of ZF to be consistently true in the universe; but if, on an occasion, we do require a consistent interpretation of ZF, we know how to obtain this too. The universe of sets has a consistent substructure that is a model of ZF. (Priest 2006b: 257–258)

There are a number of worries concerning this passage, and about the model-theoretic strategy in general, however. To start with, note that the collapsed model is constructed by leaving alone only a proper segment of the given model of ZF, which means that even if we grant that the model in question faithfully captures the structure of the cumulative hierarchy, the construction does not give us any reason to think that the *entire* hierarchy is an inner model of the naïve universe of sets.

Priest (2017: pp. 101–103) uses a different construction, due to Joel Hamkins, which avoids this problem. Rather than using a partition $\sim$, the construction uses an arbitrary covering. Without going into the details, the new construction gives a model of ZF and naïve set theory which leaves alone the entire original model but adds some extra inconsistent sets to it. The main drawback is that the new construction no longer delivers a model of NLP$_=$ but only a model of NLP, so identity needs to be defined. As we saw when discussing the material strategy, this means that identity will behave non-standardly in the theory. Priest (2017: 102) writes that 'since the name of the game at this point is recapturing the theorems of ZF, this does not matter'. But as we saw in Section 4.3, this does matter for the overall acceptability of naïve set theory.

A more serious issue is that although with the help of the Collapsing Lemma we can show that there is a model of ZF which is also a model of NLP$_=$, it does not follow from that that 'the [naïve] universe of sets is a model of ZF'. For this, we would need the assurance that one of the models obtained by using the Collapsing Lemma faithfully captures the structure of the full naïve universe of sets. But nothing we have said so far guarantees that this is the case. In his discussion, Priest says that since there are models which are models of NLP$_=$ and ZF and contain the cumulative hierarchy up to the first inaccessible, '[w]e may therefore suppose that the true interpretation of the language of set theory has these properties. This is an appealing picture' (2006b: 257). Appealing as this picture may be, it does not constitute an argument for thinking that the naïve universe of sets properly extends the cumulative hierarchy and that its structure is captured by one of the models obtained by applying the Collapsing Lemma to one of the models of ZF (see Incurvati 2010; Meadows 2015).[18]

More recently, Priest (2017: 107–108) has clarified his position and claimed that we may suppose that the universe of sets is a model of ZF '[b]ecause doing so allows one to give an account of the Cantorian picture of sets. Unless set theory

---

[18] My criticism differs from, but is somewhat related to Weir's objection to Priest that 'all the dialetheist's ZFC models are unintended in the sense that they do not capture anything like the full structure of the naïve universe of sets' (2004: 398). Weir's criticism invites the charge of being question-begging, since '[t]he thesis is precisely that one of these models does capture the full structure of the universe of sets' (Priest 2006b: 257–258, fn. 20). Although it *is* question-begging to say that all the dialetheist's models are unintended, it is also true that one cannot simply *assume* that one of the models obtained by making use of the Collapsing Lemma captures the full structure of the naïve universe of sets.

is to be unacceptably revisionist, any account must do this'. That is, the model-theoretic constructions showing that there are models of naïve set theory which are also models of ZF do not guarantee that the full naïve universe of sets is a model of ZF, but we may suppose that it is because this supposition is very fruitful in its consequences. Note that this makes the justification for the adoption of the naïve conception of set much closer to the inference to the best conception approach that we have been advocating than to Priest's (2006b: ch. 2) idea that we should endorse it because it faithfully captures our 'intuitive notion of set'.

The model-theoretic constructions showing that there are models of NLP$_=$ which arc also models of ZF are officially carried out in ZF itself (Priest 2006b: 257–258; Priest 2017: 107). On the basis of this, Meadows has objected that the model-theoretic strategy is circular, since it attempts to justify the use of ZF using that very same theory. Priest has replied that the model-theoretic constructions are not intended to *justify* ZF. They are to be conceived of as a proof (or indeed proofs) of concept that there are structures satisfying NLP$_=$ as well as ZF. Whilst the proof of concept does still employ ZF, the idea is that the use of ZF *is* justified from a paraconsistent perspective, not on the basis of the model-theoretic constructions but on the basis of the fact that supposing that the universe of sets satisfies ZF is fruitful in its consequences.

Recall, however, that it is not just the truth of the ZF axioms that the paraconsistent set theorist wants to justify, but the legitimacy of using classical logic when deriving things from them. This means, in particular, that the paraconsistent set theorist seems to be committed to the idea that she can suppose that ZF is consistent on the basis of the fruitfulness of this assumption. But this does not seem something that can just be supposed because of its fruitfulness. Returning to the analogy with the case of the intuitionist logician is helpful. The intuitionist logician would not seem to be justified in supposing that a certain domain is decidable simply because doing so is fruitful. Typically, she will instead offer an intuitionistically acceptable proof of decidability for the relevant domain.

The issue of the consistency of ZF is also connected with a related issue for the naïve set theorist. We pointed out in Chapter 1 that conceptions of set may provide reasons for the consistency of the set theory they sanction. This is because consistency is important for the classical set theorist. And although the dialetheist recognizes consistency to be a virtue, it is, according to her, one which in the case of naïve set theory is trumped by other virtues.

Nonetheless, as observed in Section 4.2, the paraconsistent set theorist still takes it as a constraint on a set theory that it should be non-trivial. But what reasons are there for thinking that naïve set theory is indeed non-trivial? It might seem that the model-constructions offered by Priest might help here, since they tell us that if there is a classical model of ZF, then there is a model of ZF and $NLP_=$, that is that if ZF is consistent, then $NLP_=$ is not trivial. But, obviously, to get from this to the non-triviality of $NLP_=$, we need the assumption that ZF is consistent. The defender of the iterative conception can make this assumption: according to her, ZF is consistent because we have an intuitive model for it in the cumulative hierarchy.[19] But how does the paraconsistent set theoriest know, or why does she believe that?[20] It cannot be because she thinks that we have an intuitive model for it in the cumulative hierarchy, since for her the existence of an intuitive model does not imply consistency, which is why she thinks that we can construct a model of NLP capturing the structure of the full universe of sets given a model of ZF. Nor can it be because she thinks that the axioms of ZF are true in the naïve universe of sets, since for the dialetheist truth does not imply consistency either. Indeed, as Priest (2017: 98) points out, once inconsistent sets enter the picture, it can be shown that if ZF's Separation Schema is true, it is both true and false.

What this brings to light is that the paraconsistent set theorist needs, after all, to say more about what the universe of sets looks like. It is not enough to simply suppose that it contains some paradigmatic inconsistent sets and has the cumulative hierarchy as an inner model.

4.6 Conclusion

I have examined three strategies for developing a paraconsistent set theory. The first strategy, the material strategy, consisted in adopting the logic LP and taking the conditional in the Axioms of Comprehension and Extensionality to be the material conditional. This strategy leads to a set theory which is very weak and whose little strength is due to the fact that we are defining identity in such a way that the direction of Leibniz's Law which almost everyone accepts fails. Attempts to consider cognate systems of LP do not fare any better.

The second strategy, the relevant strategy, consisted in taking the conditional in the axioms of naïve set theory to be a depth relevant conditional. I argued that the

---

[19] I am here setting aside the question of whether the Axiom of Replacement is justified on the iterative conception. The discussion can be easily rephrased in terms of Z^+.

[20] This criticism of Priest's strategy was inspired by Dummett's analogous objection to Field's strategy for rehabilitating nominalism in the philosophy of mathematics, which rests on the assumption that ZF is consistent (see Field 1980). Dummett (1991: 313) writes: 'How does Field know, or why does he believe, ZF to be consistent? Most people do, indeed; but then most people are not nominalists. [...] Our primordial reason for supposing ZF to be consistent lies in our belief that we have an intuitive model for it, the cumulative hierarchy in which the sets of rank $\alpha + 1$ comprise "all" sets of elements of rank α (together with the elements of rank α, it is necessary to add when we start with Urelemente)'.

motivations offered by Priest for adopting this conditional are not very convincing and that, in any event, the resulting set theory is very weak. True, one might try to strengthen the conditional to obtain a stronger set theory; but, in the absence of a motivation for the stronger conditional, it is not clear what the significance of the results thereby obtained would be. In particular, Weber's proposal to simply adopt the strongest logic which does not lead to triviality in the presence of Comprehension leads to an embarrassment of riches.

The third and final strategy, the model-theoretic strategy, consisted in arguing that the universe of sets models the ZF axioms and properly extends the cumulative hierarchy of sets. I argued that this strategy has several problems, chief among which is the fact that it ultimately needs to assume that the cumulative hierarchy is a consistent domain. In the presence of the model-theoretic constructions provided by Priest, this assumption would also help the paraconsistent set theorist offer reasons for thinking that naïve set theory is non-trivial. I argued, however, that the dialetheist does not seem to have any reason to believe that the cumulative hierarchy is a consistent domain.

Of course, there might be other strategies for developing a paraconsistent set theory based on the naïve conception. However, the three strategies at least exhaust the logical resources usually advocated by Priest. For, according to him, we should use LP to characterize the extensional connectives, a modal semantics including impossible worlds to characterize a depth relevant conditional, and classical logic to reason over consistent domains. The three strategies we explored correspond precisely to these three kinds of logical resources. Hence, the prospects for developing a paraconsistent set theory in keeping with Priest's dialetheist program look bleak.

5

The Limitation of Size Conception

One size does not fit all.

Proverb, early seventeenth century

In Chapter 4, we explored the possibility of *rehabilitating* the naïve conception of set by revising the logic of set theory: we hold on to the thought that every condition determines a set, but avoid triviality by abandoning some classically valid inferences.

In this and the next chapter we will look at two attempts to *modify* the naïve conception. The thought is that the core idea of the naïve conception – that there is an intimate connection between sets and properties – can be preserved, even in the presence of classical logic, as long as we incorporate into the conception the idea that certain properties are pathological and, for this reason, do not determine a set.

Our main task in this chapter is to discuss the idea that the pathological properties are those that apply to too many things. Before turning to this task, however, we shall consider an idea that might occur to one quite naturally. This is the idea that those properties are pathological that give rise to inconsistency.

5.1 Consistency Maxims

The view that those properties are pathological which give rise to inconsistency – that is to say, the Naïve Comprehension Schema should be restricted according to consistency maxims – can be traced back to Quine. He argued that the naïve conception of set is the only intuitive conception of set that we have. Alas, it is inconsistent.

But a striking circumstance is that none of these proposals, type theory included, has an intuitive foundation. None has the backing of common sense. Common sense is bankrupt, for it wound up in contradiction. Deprived of his tradition, the logician has had to resort to myth making. That myth will be best that engenders a form of logic most convenient for mathematics and the sciences; and perhaps it will become the common sense of another generation. (Quine 1941: 27)

Faced with the inconsistency of the Naïve Comprehension Schema and the conception of set lying behind it, the logician should not attempt to find an alternative, consistent conception of set.[1] Her task is instead to find a way of *restricting* the Comprehension Schema. However, in keeping with the maxim of minimum mutilation, this schema should be restricted as little as possible so as to avoid the paradoxes.[2]

Only because of Russell's paradox and the like do we not adhere to the naive and unrestricted comprehension schema [...] Having to cut back because of the paradoxes, we are well advised to mutilate no more than what may fairly be seen as responsible for the paradoxes. (Quine 1963: 50–51)

It is a short step from this to the idea of restricting the Naïve Comprehension Schema according to consistency maxims. For it is natural to think that the instances of the Comprehension Schema responsible for the paradoxes are precisely the ones that give rise to inconsistency.

Quine's remarks appear to be motivated by what has come to be called his *anti-exceptionalism* about logic and mathematics, famously expounded in Quine 1951b.[3] On this view, logical and mathematical theories should be assessed and evaluated in the same way as empirical scientific theories, that is, on the basis of adequacy to the data as well as how well they fare with respect to theoretical virtues such as simplicity, strength, elegance and unity. And, in line with this, the maxim of minimum mutilation, which we apply when revising the total system in the light of recalcitrant experience, is to be applied when revising our logical and mathematical theories in the light of the paradoxes.

Perhaps more surprising is the fact that Zermelo, in the paper in which he gave the first axiomatization of set theory, describes the methodology he used to formulate his axioms as being remarkably similar to the one advocated by Quine:

---

[1] As Decock (2002: 236, fn. 295) observes, there is some irony here, since it would not be preposterous to suggest that the iterative conception, so fiercely opposed by Quine, has become the common sense of subsequent generations. And, to double up on the irony, main exponents of the iterative myth have criticized the set theory advanced by Quine for not being intuitive. More on this in the next chapter.

[2] The maxim of minimum mutilation says that we ought 'to disturb the total system as little as possible' when revising it in the light of recalcitrant experience (Quine 1951b: 41). It features throughout Quine's writings, but the actual label first appears in Quine 1970.

[3] Williamson (2007) uses the term 'anti-exceptionalism' to describe his approach to philosophical methodology, which he also extends to logic and mathematics.

In particular, in view of the '*Russell* antinomy' of the set of all sets that do not contain themselves as elements, it no longer seems admissible today to assign to an arbitrary logically definable notion a 'set', or 'class', as its 'extension'. Cantor's original definition of a 'set' as 'a collection, gathered into a whole, of certain well-distinguished objects of our perception or our thought' therefore certainly requires some restriction; it has not, however, been successfully replaced by one that is just as simple and does not give rise to such reservations. Under these circumstances there is at this point nothing left for us to do but to proceed in the opposite direction and, starting from 'set theory' as it is historically given, to seek out the principles required for establishing the foundations of this mathematical discipline. In solving the problem we must, on the one hand, restrict these principles sufficiently to exclude all contradictions and, on the other, take them sufficiently wide to retain all that is valuable in this theory. (Zermelo 1908: 189–191)

Maddy (1988: 485) argued that Zermelo saw himself as attempting to deliver as much as possible of the Naïve Comprehension Schema whilst steering clear from the paradoxes. With Zermelo, Maddy suggests, emerged the 'one step back from disaster' rule of thumb: if a natural principle leads to contradiction, the principle should be weakened just enough to block the contradiction.

   The proposal that Comprehension should be restricted to those instances that do not give rise to inconsistency can be understood according to a distributive and a collective reading of 'give rise to inconsistency'. Following Incurvati and Murzi 2017, we now show that, on either reading, the proposal does not work.

The Distributive Reading: Local Consistency

According to the distributive reading, the Naïve Comprehension schema should be restricted to those instances that do not give rise to inconsistency *on their own*. The idea is to take an instance τ of (Comp) to be admissible just in case it is consistent, in the sense that τ does not imply a contradiction in first-order logic.

   However, this restriction does not suffice to obtain a consistent theory, for the simple reason that there are instances of (Comp) which are *individually* consistent – or, at least, they can be shown to be such given minimal assumptions – but *jointly* inconsistent. Two examples of this phenomenon are given in Incurvati and Murzi 2017.

   The first example makes use of the following instances of (Comp):

(1) $\exists y \forall x (x \in y \leftrightarrow x = x)$,
(2) $\forall z \exists y \forall x (x \in y \leftrightarrow (x \in z \,\&\, x \notin x))$.

(1) asserts the existence of a universal set. It is clearly consistent, since it is true in a model consisting of a looped one-node graph, in which variables range over nodes, and in which $x \in y$ just in case there is a directed edge from the node assigned to

y to the node assigned to x.[4] (2) asserts, for any set z, the existence of the set of all things in z that are not members of themselves. It is also clearly consistent, since it is true in any model consisting of a single hereditarily finite set,[5] such as the model consisting of just one object a such that $a \notin a$. However, it is routine to derive from (1) and (2) together the existence of the Russell set and hence a contradiction. Note that (2) is, in effect, an instance of ZF's Separation Schema. Indeed, the same argument we have used to derive a contradiction is used in iterative set theory to prove that there is no universal set given that there is no Russell set.

(2) is an instance of (Comp) with *parameters*.[6] In the light of this, one might try to dispense with the problem by banning instances of this kind (as well as ones that are individually consistent).

The second example shows that this will not work. The example makes use of the following observation, which goes back to Boolos (1993: p. 233):

Lemma 5.1.1. *For each φ in $\mathcal{L}_\in$,*

(i) $\exists y \forall x (x \notin y) \vdash \varphi \rightarrow \exists y \forall x (x \in y \leftrightarrow (\neg\varphi \ \& \ x \notin x))$,
(ii) $\vdash \exists y \forall x (x \in y \leftrightarrow (\neg\varphi \ \& \ x \notin x)) \rightarrow \varphi$.

Now consider the following instances of (Comp):

(3) $\exists y \forall x (x \in y \leftrightarrow (\neg\sigma \ \& \ x \notin x))$,
(4) $\exists y \forall x (x \in y \leftrightarrow (\sigma \ \& \ x \notin x))$,

where σ is a sentence such that both it and its negation are consistent. (For instance, σ could be the first-order formalization of 'There are exactly three objects'.) The first part of Lemma 5.1.1 tells us that, in the presence of the Empty Set Axiom, we can turn any proof of a contradiction from (3) into a proof of a contradiction from σ, and any proof of a contradiction from (4) into a proof of a contradiction from $\neg\sigma$. Hence, (3) and (4) are individually consistent. However, they are jointly inconsistent, since, by the second part of Lemma 5.1.1, (3) entails σ, and (4) entails $\neg\sigma$.

The Collective Reading: Global Consistency

The theory obtained by restricting the Naïve Comprehension Schema to instances which do not give rise to inconsistency on their own is inconsistent. For there

---

[4] In fact, if one is happy to countenance non-well-founded sets, an even simpler model will do, namely the one consisting solely of a set a such that $a = \{a\}$. We will explore theories admitting of such sets in the next two chapters.

[5] A set is said to be *hereditarily finite* iff it is finite, its members are finite, the members of its members are finite, and so on. In other words, a set is hereditarily finite iff its transitive closure is finite.

[6] Recall that an instance of (Comp) with parameters is one in which $\varphi(x)$ contains free variables other than x.

are pairs of such instances which do give rise to inconsistency when conjoined. The following diagnosis of what has gone wrong is tempting: the injunction to restrict the Naïve Comprehension Schema according to consistency maxims has been understood *locally*, but it ought to be understood *globally*. That is to say, the Naïve Comprehension Schema should be restricted to those instances that do not give rise to inconsistency where 'give rise to inconsistency' receives a collective reading. Formally, this is tantamount to requiring that the acceptable instances of (Comp) should form a maximally consistent set.

Indeed, the collective reading appears to better capture the spirit of Quine's and Zermelo's remarks: we are keeping as many instances of Naïve Comprehension as possible whilst avoiding the paradoxes. Even more explicitly, Emerson Mitchell, who regards himself as following a suggestion of Alonzo Church (1974), suggests basing set theory solely on the heuristic principle

of specialized comprehension, that set theory is constructed by assuming as many cases as consistently possible of the naive and inconsistent comprehension axiom scheme [. . .]. (Mitchell 1976: 1)

The collective reading of the consistency maxim is reminiscent of a proposal for the case of truth due to Horwich (1990: 42). To deal with the semantic paradoxes, he suggested restricting the Equivalence Schema

$$\text{The proposition } that \text{ } p \text{ is true iff } p \qquad\qquad \text{(E)}$$

to the maximally consistent set of its instances. However, working with sentences rather than propositions, McGee (1992) proved a theorem which spells trouble for Horwich's proposal. For McGee's Theorem shows, first, that when the base theory for the truth theory is at least as strong as Robinson Arithmetic Q,[7] any maximally consistent set of instances of the Equivalence Schema will not be recursively axiomatizable. So, it is unclear whether any such set can be said to be a *theory* of truth. And second, McGee's Theorem entails that there are several and yet incompatible maximally consistent sets of instances of the Equivalence Schema. So it is incorrect to speak of 'the' maximally consistent set of instances of the Equivalence Schema: we are confronted with an embarrassment of riches.

In Incurvati and Murzi 2017, a generalization of McGee's Theorem is given and is used to show that the situation for set theory is the same as that for truth. In Appendix 5.A, we give a precise formal statement and proof of the result. Here we focus on two key consequences of it. First, consider the instance of (Comp) asserting the existence of the empty set, that is

---

[7] Essentially, Q is PA without the Axiom Schema of Induction.

$$\exists y \forall x (x \in y \leftrightarrow x \neq x).$$ (Comp$_\emptyset$)

The result in Incurvati and Murzi 2017 tells us that any theory which comprises a maximally consistent set of instances of (Comp) including (Comp$_\emptyset$) is negation complete. Hence, if it interprets Q, such a theory cannot be recursively axiomatized. But (Comp$_\emptyset$) should be an acceptable instance of (Comp) if any is.[8] And set theory ought to interpret Q if it is to have the roles traditionally ascribed to it, such as those of being a theory of the infinite and a foundation for mathematics. Thus, no reasonably strong recursively axiomatizable theory of sets can be obtained by selecting a maximally consistent set of instances of (Comp): similarly to the case of truth, any such set can hardly be regarded as a set *theory*.

Second, it follows from the result in Incurvati and Murzi 2017 that for any consistent sentence σ, there is a maximally consistent set of instances of (Comp) implying σ, and another one implying $\neg\sigma$.[9] There are several, *mutually incompatible* maximally consistent sets of instances of the Naïve Comprehension Schema, and the injunction to restrict the Schema according to consistency maxim gives us no indication as to how to choose between them. Just as in the case of truth, we are confronted with an embarrassment of riches.

The upshot is that if we want to construct our theory of sets by restricting the Naïve Comprehension Schema in some way, we need to be guided by more than the mere desire to maintain as many instances of it as we possibly can. Some instances of the Schema will have to be banned not because they give rise to inconsistency, but because of some other feature that renders them unacceptable even in cases in which it does not result in inconsistency.[10] In this and the next chapter we shall discuss two classes of attempts to locate this feature: semantic and syntactic respectively.

---

[8] Admittedly, some philosophers have had qualms about the existence of the empty set (see Oliver and Smiley 2006 and references contained therein). But such metaphysical scruples can be sidestepped here, since all that is really needed is the existence of some set – be it empty or not.

[9] Indeed, the proof in Appendix 5.A establishes more than the mere existence of the relevant maximally consistent sets of instances: it also provides a step-by-step procedure for *constructing* them.

[10] The diagnosis of the set-theoretic paradoxes given in Boolos 1993 offers a different route to a similar conclusion. On that diagnosis, the paradoxes arise because whilst by Cantor's Theorem there are strictly more properties of objects over a given domain than objects, Naïve Comprehension demands that there be at least as many sets as there are properties. It would then seem that to develop a coherent conception of set it will suffice to restrict the domain of properties to which there corresponds a set not assigned to any other non-coextensive property. However, a generalization of the Zermelo-König inequality – provable in ZFC and stating that the cofinality of $|\mathcal{P}(a)|$ is strictly larger than $|a|$ – suggests that any such assignment must leave out at least as many properties as there are properties altogether. And this, in turn, suggests that there are as many such assignments as there are properties altogether. See Uzquiano 2015b: 339 for details.

5.2 Cantor Limitation of Size

The idea that the pathological properties are those that apply to too many things –
and that, as a result, they do not determine a set – can be traced back to Cantor. In
Chapter 1, we quoted a passage from a 1899 letter to Dedekind where Cantor tells us
what he means by a 'set' in terms of the notion of consistency. That passage is part
of a longer one in which Cantor distinguishes between *consistent* and *inconsistent*
multiplicities. He writes:

If we start from the notion of a definite multiplicity (a system, a totality) of things, it is
necessary, as I discovered, to distinguish two kinds of multiplicities (by this I always mean
definite multiplicities).

For a multiplicity can be such that the assumption that *all* of its elements 'are together'
leads to a contradiction, so that it is impossible to conceive of the multiplicity as a
unity, as 'one finished thing'. Such multiplicities I call *absolutely infinite* or *inconsistent*
multiplicities.

As we can readily see, the 'totality of everything thinkable', for example, is such a
multiplicity; later still other examples will turn up.

If on the other hand the totality of the elements of a multiplicity can be thought of
without contradiction as 'being together', so that they can be gathered together into '*one*
thing', I call it a *consistent multiplicity* or a 'set'. (Cantor 1899: 114)

What Cantor calls a 'multiplicity' (*Vielheit*) seems to be what in contemporary
parlance is referred to as a 'plurality'. However, for uniformity with the rest of
this chapter, we will frame the discussion of Cantor's view in terms of properties,
but nothing will hinge on this, and everything we shall say can be easily cast in
terms of pluralities.

Given this proviso, we can see Cantor as suggesting that there are two kinds of
properties – those which determine a set and those which do not. But when does a
property succeed in determining a set? In the quoted passage, Cantor seems to be
saying that a property determines a set just in case the assumption that it does does
not lead to contradiction. It is important to get clear about what Cantor might have
in mind here.

On one interpretation, Cantor is saying that a property determines a set just in
case the assumption that it does does not lead to contradiction in classical logic.
But if this interpretation is correct, Cantor would be committed to accepting all
instances of the Naïve Comprehension Schema which do not lead to contradiction
in classical logic. And this, as we saw in the previous section, leads to inconsistency.

However, there are reasons to think that this interpretation is not correct. For Cantor mentions as an example of an inconsistent multiplicity the totality of everything
thinkable; another example of a multiplicity that he took to be inconsistent is that

of the ordinal numbers. And as Hallett (1984: 167) notes, we know (and will see in the next chapter in more detail) that whilst assuming the existence of the set of all non-self-membered sets leads to a contradiction in classical logic, the assumption that there exists a set of all things (whether thinkable or not) or a set of all ordinal numbers leads to a contradiction only in the presence of some set-theoretical assumptions – typically, the Separation Axiom. For instance, using Separation we can prove Cantor's Theorem, which allows us to derive a contradiction from the existence of the universal set (see Section 1.B).

Thus, Cantor's suggestion seems to be that a property determines a set just in case the assumption that it does does not lead to contradiction *using unobjectionable mathematical reasoning*, taking it for granted that the Axiom of Separation is unobjectionable.

The notion of a contradiction that can be derived with the help of unobjectionable mathematical reasoning is far from transparent, however. One symptom of this is that this notion appears to provide us with little guidance on how to identify the problematic properties. Cantor, however, goes further and provides an account of a feature that promises to do just this. The idea is that the property On of being an ordinal provides a yardstick for non-sethood. To state the principle Cantor subscribed to, it will be helpful to introduce some terminology and notation. We take the opportunity to introduce some other notation and terminology to be used in the remainder of the chapter. If anything that is F is also G we say that F is a *subproperty* of G and write it $F \sqsubseteq G$. We say that F *is as big as G* (in symbols: $F \cong G$) if there is a second-order relation between the Fs and the Gs which is a one-to-one correspondence. Moreover, we say that F *is at least as big as G* (in symbols: $G \preceq F$) if there is some $H \sqsubseteq G$ such that $H \cong F$. Finally, we say that F *is smaller than G* (in symbols: $F \prec G$) if $F \preceq G$ but $G \npreceq F$. We can now state the principle Cantor subscribed to:

Cantor Limitation of Size. F determines a set iff On $\npreceq F$.

Why think that Cantor endorsed this principle? As Hallett's (1984: 168–169) points out, the left-to right direction follows from three claims Cantor makes. The first is that the 'system Ω of all numbers [the system of all ordinals] is an inconsistent, absolutely infinite multiplicity' (Cantor 1899: 115).

Bigness of On. On does not determine a set.

The second is that '[t]wo equinumerous multiplicities either are both "sets" or are both inconsistent' (Cantor 1899: 114). This is a form of Replacement for multiplicities, and in our terminology becomes

> **Property Size Replacement.** If $F \cong G$, then either both F and
> G determine a set or neither of them does.

The third is that 'every submultiplicity of a set is a set' (Cantor 1899: 114). This is
a form of Separation for multiplicities, which using our terminology reads:

> **Property Separation.** If F determines a set and $G \sqsubseteq F$, then G
> determines a set.

We can then show:

> **Proposition 5.2.1.** *Assume Bigness of* On, *Property Size Replacement and Property
> Separation. Then the left-to-right direction of Cantor Limitation of Size holds.*

Proof. Suppose that On $\preceq F$. Then there is a G such that $G \sqsubseteq F$ and $G \cong$ On.
By Bigness of On and Property Size Replacement, G does not determine a set. But
then, by Property Separation, F does not determine a set either. $\qquad\square$

But not only was Cantor committed to principles that imply the left-to-right direction of Cantor Limitation of Size; he made explicit and frequent use of it: in order to
establish that some given multiplicity is not a set, he would show that the ordinals
can be embedded into it.

In the 1899 letter to Dedekind, for instance, Cantor gives a sketch of the series
of alephs. He then asks whether there is a set whose cardinality is not an aleph. His
answer is negative:

> If we take a definite multiplicity V and assume that *no aleph* corresponds to it *as its
> cardinal number*, we conclude that V must be *inconsistent*. For we readily see that, on
> the assumption made, the whole system Ω is projectible into the multiplicity V, that is
> there must exist a submultiplicity V' of V that is equinumerous to the system Ω.
>
> V' is inconsistent because Ω is, and the same must therefore be asserted of V.
>
> Accordingly, every transfinite *consistent multiplicity*, that is, every transfinite set, must have
> a *definite aleph* as its cardinal number. (Cantor 1899: 116–117)

In this passage Cantor assumes the left-to-right direction of Cantor Limitation of
Size. But he also claims that if F has no aleph as a cardinal number, then there
must be some subproperty G of F such that $G \cong$ On. Putting this together with
his conclusion that every set has a cardinal number which is an aleph, we see that
Cantor is also committed to the right-to-left direction of Cantor Limitation of Size,
stating that, given some F, if there is no subproperty G of F such that $G \cong$ On,
then F determines a set (see Hallett 1984: 171–172).

Thus, we can conclude that Cantor provided a criterion for when a property
determines a set. The criterion tells us when a property determines a set on the basis
of its *size*: it tells us that a property determines a set if and only if it is not as big

as or bigger than the property of being an ordinal. It is not implausible, therefore, to think that Cantor's 1899 letter to Dedekind signals the birth of the *limitation of size* idea:

> **Limitation of Size Idea.** A property determines a set iff it is not too big.

One central question for any account based on the limitation of size idea is: 'How big is too big?'. Cantor's answer is that the ordinals provide a measure of how big is too big: if we can embed the ordinals into some things to which the property F applies, then F is too big to determine a set.

Cantor, then, holds that On does not determine a set, and uses this fact to characterize sethood. However, the question arises as to why On *itself* does not determine a set. We will return to this question in Section 5.8. For his part, Cantor does not give an explicit answer to this question, but seems to suggest that On does not determine a set because the assumption that it does leads to the Burali-Forti Paradox. In fact, having explained the distinction between consistent and inconsistent multiplicities, Cantor goes on to show that taking the ordinals to form a set leads to contradiction (see Cantor 1899: 115). Thus, his view seems to be that a property determines a set only if the supposition that it does does not lead to contradiction, and we can provide a criterion for establishing which properties are such that the supposition that they determine a set leads to contradiction: they are the properties which are at least as big as On, and we know that the supposition that this property determines a set leads to contradiction.

The term 'limitation of size' first appears in Russell 1906. According to Russell, the paradoxes show that not every condition determines a set or, as he puts it, that not every propositional function determines a class. But when does a condition determine a set? Russell considers three possible approaches. One approach is to say that no propositional function determines a class – there are no classes and talk of classes is just a 'convenient abbreviation' (Russell 1906: 37). This gives rise to what Russell calls the *no class theory*. The remaining two approaches do take classes to exist but place restrictions on which propositional functions determine a class – which propositional functions are *predicative*, as Russell puts it. One of these two approaches takes a propositional function to be predicative if its syntactic expression is simple. This gives rise to the *zigzag theory*. We shall return to this theory in the next chapter, since it can be reasonably considered a precursor of the set theories we shall discuss there. The other of these two approaches takes a propositional function to be predicative if the size of the class it determines is not 'excessive'. This is the *theory of limitation of size*, and Russell tells us that

[t]his theory naturally becomes particularized into the theory that a proper class must always be capable of being arranged in a well-ordered series ordinally similar to a segment of the series of ordinals in order of magnitude; this particular limitation being chosen so as to avoid the Burali-Forti contradiction. (Russell 1906: 43)

Thus, like Cantor, Russell understands the limitation of size idea in terms of Cantor Limitation of Size, and takes the specific choice of On as a yardstick of excessive bigness to be motivated by the desire to avoid the Burali-Forti contradiction.

5.3 Ordinal Limitation of Size

Cantor Limitation of Size states that a property F determines a set if and only if On is not as big as F. This might seem to be equivalent to the following principle, stating that a property determines a set just in case it is smaller than the property of being an ordinal:

>   **Ordinal Limitation of Size.** F determines a set iff $F \prec$ On.

However, the left-to-right direction of Cantor Limitation of Size is, on the face of it, weaker than the right-to-left direction of Ordinal Limitation of Size: the former, but not the latter, seems compatible with a property determining a set and yet not being smaller than the property of being an ordinal. What is needed to bridge the gap, it seems, is a principle telling us that properties are always comparable in size:

>   **Property Comparability.** For any F and G, we have that $F \prec G$
>   or $F \cong G$ or $G \prec F$.

Whilst it is known that cardinal comparability for *sets* – that is, the principle that all cardinals are comparable – is equivalent to the Axiom of Choice (over very weak set theories), less is known about the strength of Property Comparability compared to that of other second-order choice principles. In particular, in pure second-order logic, the Global Well-Ordering Principle – that is the second-order statement that the universe can be well-ordered – implies Property Comparability, and Property Comparability implies the second-order version of the Axiom of Choice. However, it is not known whether these two implications reverse, although at most one of them can reverse (Shapiro 1991: 106–108). By contrast, if we move to the ZFC_2 context, these second-order choice principles become equivalent.

   Linnebo (2010: 162) has shown that as well as bridging the gap between Cantor Limitation of Size and Ordinal Limitation of Size, Property Comparability actually enables us to *prove* Ordinal Limitation of Size solely in the presence of *Property Replacement*, stating that if F determines a set and $G \preceq F$, then G determines a set:

Theorem 5.3.1 (Linnebo). *Property Comparability and Property Replacement jointly imply Ordinal Limitation of Size.*

Sketch of proof. Suppose Property Comparability and Property Replacement hold. We want to prove Ordinal Limitation of Size. We first establish the left-to-right direction. To this end, observe that Property Replacement easily implies Property Size Replacement and, via Burali-Forti reasoning, that On does not determine a set. Moreover, it is a well-known fact that Property Replacement implies Property Separation. By Proposition 5.2.1, we then have that the left-to-right direction of Cantor Limitation of Size holds. The desired conclusion follows by Property Comparability.

To establish the right-to-left direction of Ordinal Limitation of Size, suppose $F \prec$ On. Then, we can embed the ordinals into the Fs and thereby obtain a well-ordering of F. Such a well-ordering is either isomorphic to the members of some ordinal or to the ordinals themselves. If the latter, then $F \nprec$ On, contradicting the initial supposition. Hence, the former must be the case, and we can use Property Replacement to conclude that F determines a set. $\square$

In any event, Ordinal Limitation of Size articulates a version of the limitation of size conception which appears to be faithful to Cantor's approach. For Cantor made explicit use of reasoning involving arbitrary choices, and **AC** in conjunction with Cantor Limitation of Size yields Global Well-Ordering, which implies Property Comparability, thus delivering Ordinal Limitation of Size.

Moreover, Ordinal Limitation of Size is probably what a defender of a limitation of size approach based on ordinals will want to settle on, since she is unlikely to want a property to determine a set solely on the grounds that it or some sub-property thereof is not comparable in size with the property of being an ordinal. In other words, once size is the criterion for determining whether a property determines a set, Property Comparability becomes an attractive principle, which provides – via Theorem 5.3.1 – a reason for focusing on Ordinal Limitation of Size.

5.4 Von Neumann Limitation of Size

According to the limitation of size idea, a property determines a set just in case it is not too big. Cantor understood 'too big' by reference to the ordinal number series. But there are other ways of understanding it. One such way first appeared in the work of von Neumann (1925), who presented another instance of the limitation of size idea.

Like Cantor, von Neumann takes a certain property to be the paradigmatic case of a big property, and uses it to formulate a criterion for saying when *any* property is too big. However, the property von Neumann focuses on is not the property of being an ordinal, as in Cantor's case, but the property which applies to everything: the universal property $[x : x = x]$, which we abbreviate with U. A limitation of size principle inspired by von Neumann would then say that a property determines a set just in case it is not as big as U. That is:

Von Neumann Limitation of Size. F determines a set iff $F \not\approx U$.

Three remarks about Von Neumann Limitation of Size are in order. First, it is a very strong principle: in combination with the Burali-Forti reasoning, it delivers Global Well-Ordering and hence Property Comparability.

Second, the principle is only *inspired* by von Neumann's work. In a letter to Zermelo, von Neumann (1923) actually states his limitation of size principle as follows:

A set is 'too big' iff it is equinumerous with the set of all things.

Thus, for von Neumann it is not as if some properties do not determine a set because they are too big. Rather, every property determines a set, but some sets are 'too big' and, because of this fact, behave differently. In fact, as we know, they *need* to behave differently, otherwise paradox will ensue. What is the difference in their behaviour, then? According to von Neumann, sets that are too big cannot be members of other sets, and this is meant to block the paradoxes. Bernays later recast von Neumann's distinction between sets that are and sets that are not too big in terms of the distinction between sets and proper classes.

Third, having first introduced his limitation of size principle informally (1925: 397–398), von Neumann (1925: 400) goes on to incorporate his limitation of size principle as an *axiom* (Axiom IV.2) in a formal system which led to the development of NBG.[11] In fact, von Neumann's system contains an axiom that effectively amounts to the assertion that those sets which cannot be members of other sets are exactly those that can be put in one-to-one correspondence with the set of all sets.

Von Neumann's suggestion makes clear that we can distinguish between two types of limitation of size of principles, *maximalist* and *non-maximalist* ones. A maximalist understanding of the limitation of size idea equates 'too big' with 'as

---

[11] Von Neumann's axiomatization is formulated in terms of functions and arguments, so his Axiom IV.2 is also formulated in this manner.

big as the universal property', thereby giving rise to Von Neumann Limitation of Size. A non-maximalist understanding of the limitation of size idea, on the other hand, equates 'too big' with 'as big as or bigger than F', where F is not simply U. Assuming again Property Comparability in the background, the non-maximalist understanding of limitation of size gives rise to the following template:

Non-Maximalist Limitation of Size. F determines a set iff $F \prec G$.

Ordinal Limitation of Size takes G to be the property of being an ordinal, but other choices are possible. What distinguishes non-maximalist versions of limitation of size from the maximalist version is that the latter guarantees that a property will determine a set if it is not as big as the universal property, whereas the non-maximalist version fails to guarantee that, since F may well be smaller than the universe. Thus, whilst non-maximalist versions of limitation of size may leave it open whether F is as big as the universe, the maximalist version, trivially, does not.[12]

5.5 Frege-Von Neumann Set Theory

So far, we have looked at various limitation of size principles, but have said little about the kind of conception of set to which they give rise. The starting point of the limitation of size conception is, in fact, the naïve conception. The limitation of size theorist takes it to be an important insight of the naïve conception that sets are determined by properties: the limitation of size conception is a *logical* conception. But she rejects the idea that every property determines a set. Certain properties, she says, are pathological and do not determine a set.

Limitation of size principles are invoked precisely to determine *which* properties are pathological. According to the limitation of size theorist, the pathological properties are the properties that are true of too many things, and this idea can be spelled out, as we have seen, in various ways. Thus, it is natural to think of the limitation of size conception as a modification of the naïve conception: we hold on to the thought that sets are determined by properties, but in so doing we make it clear that certain properties are *bad* and do not determine a set. These are the properties that are true of too many things.

---

[12] It is worth noting that in some cases non-maximalist versions of limitation of size imply the maximalist version, perhaps in the presence of some additional assumptions. For instance, in the presence of the NBG axioms, AC and Cantor Limitation of Size jointly imply Global Well-Ordering. And, as von Neumann (1925) already pointed out, Global Well-Ordering not only follows from but is equivalent to Von Neumann Limitation of Size.

Now, once the limitation of size conception is presented in this way, it suggests a natural way of formulating a set theory based in it: start from a formulation of the naïve conception and build a limitation of size principle into it. One way to proceed would be to formulate a theory which embodies the limitation of size conception by retaining the Axiom of Extensionality and modifying the Naïve Comprehension Schema. However, quantification over properties is naturally formulated using a second-order language, as we have been doing so far. This suggests using the formulation of the naïve conception due to Frege, which we encountered in Chapter 1. According to this formulation, to every concept there corresponds its extension, analogously to the way in which, according to the naïve conception, to every property there corresponds a set of all and only the things which have that property. Recall that Frege's version of the naïve conception is embodied by the following principle:

$$\mathrm{Ext}(F) = \mathrm{Ext}(G) \leftrightarrow \forall x (Fx \leftrightarrow Gx), \tag{V}$$

where $\mathrm{Ext}(F)$ is the *extension* of the concept F.

Now (V) is inconsistent, but we can try to repair it by incorporating into it a limitation of size principle. We follow in Boolos's (1989: 98–99) footsteps and formulate a different abstraction principle. This axiom retains the naïve thought that every concept F is associated with an object $\mathrm{Subt}(F)$, the *subtension* of F. But the subtensions of two concepts are now taken to be identical just in case neither is small, where smallness is understood according to Von Neumann Limitation of Size: we take a concept F to be *small* just in case $F \not\cong U$.[13] This gives rise to the following amendment of (V):

$$\mathrm{Subt}(F) = \mathrm{Subt}(G) \leftrightarrow F \approx G, \tag{New V}$$

where $F \approx G$ is to be read as 'either F and G are coextensive or neither is small'.

Boolos dubs the theory obtained by adding (New V) to standard second-order logic FN, from Frege-von Neumann, since it combines Fregean naïvete with von Neumann-type size restrictions. He points out that we have good reason to believe FN to be consistent, namely that it has a model in the natural numbers.[14] The model

---

[13] Boolos actually defines a concept F to be small just in case $F \not\preceq U$, but the two definitions are equivalent, since it can be shown that if $U \preceq F$, then $U \cong F$.

[14] Of course, the fact that FN has a model in the natural numbers does not imply (nor give us any reason to believe) that it has a model which is rich enough to sustain the reconstruction of a substantial amount of mathematics. Indeed, Shapiro and Weir (1999: 318) have shown that it is consistent with ZFC that FN has no uncountable models.

is obtained by taking concepts to be sets of natural numbers, sending each finite concept to a different natural number other than 0 and all infinite concepts to 0.[15]

What is the strength of FN? To start with, the result that FN has a model in the natural numbers is best possible, in the sense that any model of FN must be at least of the size of the natural numbers.

Theorem 5.5.1 (Boolos). FN *is only satisfiable in an infinite domain.*

Proof. Let 0 be the subtension of $[x : x \neq x]$ and $\ddagger$ be the subtension of U. Since there is at least one object, $[x : x \neq x]$ is small, and $[x : x \neq x] \not\approx U$. Hence, $0 \neq \ddagger$. Thus, there are at least two objects. Now define the successor operation s by setting $sy = \text{Subt}([x : x = y])$. It is easy to see that for every y, sy exists, since for each y, exactly one object falls under $[x : x = y]$ and so $[x : x = y]$ is small. Moreover, we have that for each y, $0 \neq sy$, since $[x : x \neq x]$ is small and $\neg \forall x (x \neq x \leftrightarrow x = y)$. Finally, sy is one-to-one, that is, if $sy = sz$, then $y = z$. For $[x : x = y]$ is small and so $\forall x (x = y \leftrightarrow x = z)$. But it is well-known that any model in which for every y the successor of y exists and is never 0, and the successor operation is one-to-one, must be infinite. □

As Boolos (1989: 99) notices, this means that arithmetic can be carried out in FN by letting $[x : x$ is a natural number$]$ be the concept under which x falls just in case it satisfies second-order induction.

To develop some set theory in FN, Boolos defines membership in the obvious way, namely by stipulating that y is a member of x just in case x is the subtension of a concept under which y falls. In symbols:

$$y \in x \leftrightarrow \exists F (x = \text{Subt}(F) \wedge Fy).$$

This is in keeping with the fact that the limitation of size conception is a logical conception: membership is derivative upon falling under a concept. Boolos then take sets to be subtensions of small concepts, and examines which of the ZFC axioms are validated by FN. We have:

Theorem 5.5.2 (Boolos). FN *implies the Axioms of Extensionality, Empty Set, Unordered Pairs, Separation, Choice and Replacement.*

---

[15] In more detail, let g be a $1 - 1$ mapping from finite subsets of ω into $\omega \backslash 0$ and define the following interpretation for the operation Subt: $\mathcal{P}(\omega) \mapsto \omega$:

$$\text{Subt}(F) = \begin{cases} g(F) & \text{if } F \text{ is finite} \\ 0 & \text{if } F \text{ is infinite} \end{cases}$$

It is easy to see that $\langle \omega, \text{Subt} \rangle$ is a model of FN, one in which a concept is small just in case it is finite.

The proofs are simple but quite involved, so we refer the reader to Boolos 1989: 98–103. Theorem 5.5.2 vindicates the oft-made claim that Separation and Replacement are justified on the limitation of size conception (see, e.g., Fraenkel et al. 1973: 501; Hallett 1984: 209; Maddy 1988: 489; Potter 2004: 229). For Separation, the idea is that if a concept F determines a set and hence is small, then any subconcept F of G will also be small and hence determine a set. For Replacement, the idea is that if a concept F determines a set and hence is small, and there is a function from F to G, then F is at least as big as G and so G is small and hence determines a set.

How about the other ZFC axioms? It follows from the fact that FN has a model in the natural numbers that Infinity is not derivable within it. Powerset is not derivable either (see Boolos 1989: 103). Moreover, if we allow *Urelemente*, neither Foundation nor Union are forthcoming. This is because the presence of the *Urelement* ‡ – the subtension of U – can be exploited to falsify them. For Foundation, take the extension of the concept $[x : x = ‡]$. This is a set, $\{‡\}$, since for every y, the concept $[x : x = y]$ is small (see the proof of Theorem 5.5.1). Now clearly, $‡ \in \{‡\}$. But since every object belongs to ‡, we also have $\{‡\} \in ‡$. For Union, just observe that the members of the sole member of $\{‡\}$ do not form a set.

However, the way in which Foundation and Union were falsified suggests that if we focus on *pure* sets, we might be able to recapture these axioms. Boolos defines the notion of a pure set in the current setting as follows. First, we say that a concept is *closed* if and only if whenever all members of a subtension fall under it, so does the subtension, that is

$$\forall x (\exists F (F = \mathrm{Subt}(F)) \wedge \forall y (y \in x \rightarrow Fy) \rightarrow Fx).$$

We then say that a set is *pure* just in case it falls under all closed concepts. With this definition on board, he goes on to show:

Theorem 5.5.3 (Boolos). FN *implies the relativizations of the Axioms of Foundation and Union to pure sets.*

Note, however, that to rule out *Urelemente* such as ‡, one need not restrict attention to pure sets (see Jané and Uzquiano 2004: 440). For say that a concept is *transitive* just in case whenever a subtension falls under it, then so do all of its members. (Transitivity in this sense, then, is a kind of converse of closedness.) A subtension is *transitive* if and only if it is the subtension of a transitive concept, and it is *hereditarily small* if and only if it falls under some transitive concept under which only small extensions fall. The idea is that hereditarily small extensions are those such that they and their transitive closure are small, and, clearly, focusing on hereditarily small subtensions suffices to dispense with ‡.

Now if rather than relativizing Foundation and Union to pure sets, we relativize them to hereditarily small subtensions, then Theorem 5.5.3 continues to hold as far as Union is concerned. But it ceases to hold with regards to Foundation. However, we do get Foundation if we further confine attention to *well-founded* hereditarily small subtensions, where a subtension is *well-founded* if it is the subtension of a well-founded concept – one under all of whose nonempty subconcepts falls some $\in$-minimal object. It can then be proved that something is a well-founded hereditarily small subtension if and only if it is a pure set.

This sheds some light on Boolos's restriction to pure sets. In restricting attention to pure sets, not only are we ruling out objects such as ‡, but we are also imposing a well-foundedness requirement on sets. Thus, the fact that the relativization of Foundation to pure sets follows from FN is not as surprising as it might seem at first sight.

5.6 The Extension of Big Properties Objection

We have understood the limitation of size conception as an attempt to capture the idea that certain properties are pathological, and fail to determine a set. However, (New V) implies that pathological properties have subtensions. In fact, they all have the same subtension, namely ‡. Since subtensions look a lot like usual, old-fashioned extensions, it seems that (New V) has committed the limitation of size theorist to accepting that pathological properties do determine sets.[16]

The obvious reply to this worry is that sets are subtensions of *small* concepts – after all, this is what Boolos take sets to be in the FN context; the subtension of the universal concept is something other than a set – a class, perhaps.

Obvious as it might seem, however, this reply is unsatisfactory (see Paseau 2007: 42). For the subtension of the universal concept is an object with members, is – unlike standard proper classes – a member of small subtensions and hence sets, and is obtained via the same operation through which sets are obtained. To deny this entity the status of set, whilst granting it to all other subtensions, looks a lot like a stipulative manoeuvre.

However, the limitation of size theorist can modify (New V) so as to avoid commitment to the extensions of big concepts. The strategy is rather natural. Rather than assigning an object to every concept, we need to formulate the abstraction axiom in such a way that it assigns an extension only to small concepts – in such

---

[16] As we saw earlier, this seems to have been von Neumann's position, who distinguished between sets that are too big and those that aren't.

a way that it is only to non-pathological properties that there corresponds a set-like object. What we need, in other words, is a principle that retains the naïve characterization of sets as extensions of concepts, but restricts it to small concepts. The following will do:

If F and G are small, then $\text{Ext}(F) = \text{Ext}(G) \leftrightarrow \forall x (Fx \leftrightarrow Gx)$.   (New V^-)

In terms of notation, we have reverted to $\text{Ext}(F)$, since in this case we are claiming that anything which is an extension is a set, as in the case of the original (V).

(New V^-) gives us what we wanted: unlike (New V), it assigns no object to the universal concept and all other big concepts. But there is a drawback. We showed earlier that FN is only satisfiable in infinite domains and hence arithmetic can be carried out within it (Theorem 5.5.1). Now let FN^- be the theory obtained by adding (New V^-) to second-order logic. Unlike FN, this theory has a model whose domain consists of exactly one object and hence does not imply the existence of infinitely many objects. The proof of Theorem 5.5.1 breaks down because the fact that for every y exactly one object falls under the concept $[x : x = y]$ does not show that this concept is small unless one can show that there are two objects. And whilst both (New V) and (New V^-) imply the existence of an object associated with the empty concept $[x : x \neq x]$, only (New V) also implies the existence of the object $\ddagger$ associated with the universal concept U.

However, the proof goes through as before once one assumes that there are at least two objects: (New V^-) implies that if there are at least two objects, there are infinitely many objects.[17] More naturally, we can obtain the same result by laying down an axiom stating that the concept $[x : x = 0]$ is small. So let FN^* be the theory obtained by adding such an axiom to FN^-. Setting aside the question of whether one can assume that $[x : x = 0]$ is small – a question to which we shall return in the next section – FN^* does not imply the existence of any object other than those associated with small concepts, but, at the same time, it does deliver some interesting set theory, as Theorems 5.5.1, 5.5.2 and 5.5.3 continue to hold if we replace FN with FN^*.

5.7 The Arbitrary Limitation Objection

Clearly, however, FN^* fails to deliver the Axioms of Powerset and Infinity, just as FN did. This is just a symptom of the fact that although the limitation of size

---

[17] Looking at things from the cardinal point of view, the situation can be clarified by noticing that a set of n elements has $2^n - 1$ proper subsets, and $2^n - 1 > n$ for $n \geq 2$.

conception maintains that a property determines a set just in case it is not too big, this tell us very little about the extent of the set-theoretic universe. For even if we assume, as Ordinal Limitation of Size and Von Neumann Limitation of Size do, that the property of being an ordinal and the universal property *are* too big, this is compatible with the universe itself being rather small, by normal set-theoretic standards.

This feature of limitation of size was noted early on by Russell, who took it to be a decisive shortcoming of the conception. Speaking of, in effect, Ordinal Limitation of Size, he writes:

A great difficulty of this theory is that it does not tell us how far up the series of ordinals it is legitimate to go. It might happen that ω was already illegitimate: in that case all proper classes [i.e. all legitimate sets] would be finite. [...] Or it might happen that ω^2 was illegitimate, or ω^ω or ω_1 or any other ordinal having no immediate predecessor. We need further axioms before we can tell where the series begins to be illegitimate. (Russell 1906: 44)

The fact that FN and FN* do not imply the Axiom of Infinity makes vivid Russell's point: there are interpretations of 'small' according to which [$x : x$ is a natural number] is not small. Thus, Von Neumann Limitation of Size seems per se compatible with the universe not containing an infinite set.

In the quoted passage, Russell hints at the fact that we can change the situation by adding further axioms. In the case at hand, we can add to FN* an axiom to the effect that the concept of being a natural number *is* small: the resulting theory implies Infinity. Or, as Boolos (1989: 103) notices, we can obtain the Powerset Axiom by adding a principle stating that if F is small, then so is [$Ext(G) : \forall x(Gx \rightarrow Fx)$]. And, obviously, we can obtain even stronger theories by adding further smallness principles. It is often thought, however, that this kind of move is desperately ad hoc: the limitation of size account is silent over the question how far the series of ordinals goes, and the problem is not solved by simply *stipulating* that it goes as far as a certain ordinal.

Since Russell, the view that this is a serious difficulty for the limitation of size conception has become widespread. Moreover, it is argued, this difficulty does not beset the iterative conception. Thus, Potter (2004: 230) takes what is effectively Russell's objection to be a reason to favour the iterative conception over the limitation of size account: whilst 'there is little prospect that [the latter] will provide support for the axiom of infinity', the iterative conception holds better prospects for success in this regard.

However, whether one holds a minimalist account or a dependency account, the iterative conception has it that sets are arranged in a cumulative hierarchy divided into levels. These levels are indexed by the ordinals, and the question arises how far up in the series of ordinals it is legitimate to go: is it legitimate to go as far as ω? How about ω_1, or the first inaccessible ordinal? This is notoriously a central question for the iterative conception. But, clearly, it is the same difficulty that Russell observed for the limitation of size conception.

It might be replied that a natural answer can be offered by the iterative conception to the question 'How far does the iterative process go?', namely 'as far as possible'. This does not tell us much until the idea of iteration as far as possible is developed to some extent, but it has been argued by some that the idea that the hierarchy extends as far as possible is capable, ultimately, to deliver the axiom of infinity and indeed axioms of higher infinity (see Section 3.6 and Incurvati 2017).

However, the limitation of size theorist seems in a position to offer a similar answer. When asked 'How big is too big?' – or better, to pursue the analogy with the iterative conception, 'How big is the concept small?' – she can answer 'as big as possible'. Indeed, this seems to have been Cantor's position (Hallett 1984: 43): if it is consistent or coherent that a concept F is small, then F is in fact small.[18]

It is therefore a mistake to take the difficulty pointed out by Russell to be a reason for preferring the iterative conception to the limitation of size account. This, of course, does not show that the difficulty can be overcome, and the jury is out on whether and to what extent principles to the effect that the hierarchy goes as far as possible as capable of sanctioning axioms of infinity. But, to stress, it seems that the limitation of size theorist will just be able to appropriate whatever answer is given by the iterative theorist.

5.8 The No Complete Explanation of the Paradoxes Objection

The next objection to the limitation of size conception targets its explanation of the paradoxes. The limitation of size conception holds that certain properties do not determine a set because they are too big, and 'too big' is understood by reference to some property which is taken to be paradigmatically too big, for instance the ordinal property On or the universal property U. But why are *these* properties too big to form a set? The limitation of size conception, the objection has it, does not provide an answer to this equation. Instead, the objection goes, the conception simply *assumes* that the property of ordinal or the universal property do not determine a set.

---

[18] The reader might wonder why this view does not fall prey to the problems highlighted with the view according to which the naïve conception should be restricted according to consistency maxims. The answer is that, following Cantor, we are disregarding properties such that the supposition that they determine a set leads to contradiction in the presence of ordinary mathematical reasoning – chiefly, the Axiom of Separation. This makes vivid the issue of what entitles the limitation of size theorist to disregard such properties, which we will discuss in the next section.

Inspired by Cantor (see Section 5.2), one might reply that the limitation of size conception does provide the required explanation. For, the reply goes, according to the limitation of size conception, the property of being an ordinal does not determine a set because the supposition that it does leads to contradiction via the Burali-Forti Paradox. Moreover, the reply goes, the reasoning involved in deriving the said contradiction only requires the Axiom of Separation, which is justified on the limitation of size conception. Similar considerations apply to the case of the universal property.

The objector might retort that this explanation is circular. For the explanation appeals to the Axiom of Separation. And whilst it is true that this axiom is typically taken to be justified on the limitation of size conception, the objector might continue, this is because the limitation of size conception is taken to *already* incorporate some specific notion of smallness. For instance, we might take Separation to be justified on the limitation of size conception because it follows from FN, as Boolos has shown. Boolos's derivation of Separation from FN goes as follows. Suppose that a is a set, say the subtension of F. Then let G be the concept $[x : x \in a \wedge Xx]$. We have that G is a subconcept of F and therefore G is small. The subtension of G is then a set which provides the witness to the relevant instance of the Separation Axiom. However, this reasoning uses the fact that if G is a subconcept of F and F is small, then G is small. And Boolos's conclusion follows because smallness is defined specifically in the way he does – that is, using Von Neumann Limitation of Size. However, this just assumes that the universal property does not determine a set.

Thus, it seems that the limitation of size conception does need to assume that some properties do not determine a set by default. However, the circularity is only apparent: for as Boolos's proof makes clear, the derivation of Separation from FN does not need any *specific* limitation of size principle, but only Property Separation. And the limitation of size theorist has the resources of justifying Property Separation *prior* to settling on any specific yardstick for excessive size: she can simply insist that if F is small and G is a subproperty of F, then G will be small too *no matter how smallness is characterized.*

However, the limitation of size explanation of why certain paradigmatic properties do not determine a set brings to light an aspect of the conception which makes it less attractive than the iterative conception. In Section 2.7 we said that Boolos deemed the limitation of size conception not to be natural. A conception, according to Boolos (1989: 92), is natural if 'without prior knowledge or experience of sets, we can or do readily acquire [it], easily understand it when it is explained to us, and find it plausible or at least conceivably true'. And, according to him, the limitation of size conception is not natural because 'one would come to entertain it only after one's preconceptions had been sophisticated by knowledge of the set-theoretic anti-nomies, including not just Russell's Paradox, but those of Cantor and Burali-Forti as well' (1989: 92). We now see that Boolos was exactly right about this. It is only because the supposition that the property of being an ordinal determines a set leads to the Burali-Forti Paradox that the limitation of size conception takes this property *not* to determine a set, and similarly for Cantor's Paradox.

This is in contrast with the iterative conception, whose explanation of why there is no set of all ordinals or no universal set does not rest on the fact that the suppo-sition that there are such sets leads to paradox (in the presence of Separation). For instance, according to the iterative conception, there is no set of all sets because this set never appears in the cumulative hierarchy of sets, no matter how far the *set of* operation is iterated. Similarly, there is no set of all ordinals because, assuming that ordinals are construed as sets, such a set is not going to be bounded in the cumulative hierarchy.[19]

Thus, the explanation of the paradoxes offered by the iterative conception appears to be superior to the one offered by the limitation of size conception, since it does not ultimately rest on the fact that supposing that certain properties determine a set leads to paradox. However, Linnebo has put forward an argument to the effect that the iterative conception and indeed any successful explanation of why some property does not determine a set must ultimately appeal to some limitation of size principle. If successful, this argument would undermine the iterative conception's claim to provide a better explanation than the limitation of size conception of why paradigmatically big concepts do not determine a set. Speaking in terms of pluralities instead of concepts/properties, Linnebo writes:

it is hard to see how an explanation of why some pluralities fail to form sets can avoid appealing to some principle of limitation of size. For [...] two widely held assumptions entail that the dividing line between pluralities that form sets and those that do not has to be a matter of their size or cardinality. More precisely, the two assumptions entail that every plurality below some threshold cardinality forms a set, whereas every plurality at or above this threshold cardinality fails to form one. (Linnebo 2010: 151)

---

[19] If ordinals are not construed as sets, then the iterative conception provides no explanation as to why they do not form a set. But this is how it should be, since if ordinals are not construed as sets, then they can be treated as individuals and there is nothing to prevent them from forming a set. The Burali-Forti contradiction would only arise if, mistakenly, we added a principle to our theory linking ordinals to sets (see Potter 2004: 58). For a discussion of the various options available when confronted with the Burali-Forti Paradox, see Florio and Leach-Krouse 2017.

Here, Linnebo is referring to Theorem 5.3.1, whose proof we gave earlier. The theorem, recall, states that Property Replacement and Property Comparability imply Ordinal Limitation of Size. Thus, we have a way to prove a limitation of size principle from two widely accepted axioms of set theory. This is a theorem and, as such, it is beyond dispute to the extent that the logic and the assumptions it uses are beyond dispute.

Let us consider first the claim that *any* explanation of why some property does not determine a set must ultimately appeal to some limitation of size principle. We will see in the next chapter that there is a conception of set – the stratified conception – which offers a radically different explanation of the paradoxes than the iterative conception, based on rejecting indefinite extensibility rather than universality (and hence admitting big collections such as the set of all ordinals or the universal set). Clearly, this explanation is in conflict with an explanation appealing to some limitation of size principle. However, this is no threat for the conception to be considered in the next chapter, since the conception leads to the rejection of the Axiom of Separation (which gives rise to indefinite extensibility) and hence of Property Replacement used in Linnebo's Theorem.

However, things are different for the iterative conception. For whilst the status of the Axiom of Replacement on the iterative conception is controversial, it would be a remarkable result that the iterative conception's explanation of the paradoxes has to appeal to some limitation of size principle on pain of giving up Replacement. However, even if we grant that the iterative conception sanctions a second-order version of Replacement, all that follows from Linnebo's result is that the iterative conception *implies* a limitation of size principle, not that its *explanation* of why a property does not determine a set must be based on considerations of size. For what the iterative conception tells us is that certain properties do not determine a set because it is never the case that all the things to which they apply show up at some unique level of the cumulative hierarchy. This explanation does not appeal to any limitation of size principle. As it happens, this explanation of why certain properties do not determine a set implies that a property determines a set just in case it is smaller than the property of being an ordinal. But this is not the iterative explanation of why those properties do not determine a set.

Thus, the limitation of size conception's explanation of the paradoxes is less attractive than the iterative one, because it ultimately rests on the fact that supposing certain properties to determine a set leads to paradox. Following Boolos, this means that the limitation of size conception is not natural: its initial plausibility and attractiveness depends on familiarity with substantial and sophisticated set-theoretic machinery.

5.9 The Definite Conception

So far, we have considered the idea of restricting (V) by incorporating into it some limitation of size principle, as in (New V) and modifications thereof. Clearly, however, the strategy of restricting (V) to small concepts is just an instance of a general strategy of restricting it to concepts that are deemed to be *good*:

$$\text{If } F \text{ and } G \text{ are good, } \text{Ext}(F) = \text{Ext}(G) \leftrightarrow \forall x (Fx \leftrightarrow Gx). \qquad \text{(GoodV)}$$

The limitation of size conception takes the good concepts to be the small ones, but other choices are possible. The idea we consider in this section is to take the good concepts to be the *definite* ones, where a concept is definite if and only if it is not indefinitely extensible:

$$\text{If } F \text{ and } G \text{ are definite, } \text{Ext}(F) = \text{Ext}(G) \leftrightarrow \forall x (Fx \leftrightarrow Gx). \qquad \text{(DefV)}$$

This gives rise to what one may call the *definite conception of set*.[20] According to the definite conception, sets are extensions of properties, but only *definite* properties determine sets:

Definiteness Idea. A property determines a set iff it is definite.

The idea that the naïve conception should be restricted according to the definitess idea fits well with the diagnosis of the paradoxes presented in Section 1.7. For according to that diagnosis, the set-theoretic paradoxes arise because on the naïve conception we have both indefinite extensibility and universality. If one then thinks that there are indefinitely extensible concepts, it is natural to try to avoid the paradoxes by restricting the naïve conception to properties that are *not* indefinitely extensible. Indeed, Russell himself, when introducing the doctrine of limitation of size, motivates it by appealing to the existence of processes which seem essentially incapable of terminating. That is, he appeals to the existence of indefinitely extensible concepts:

The reasons for recommending this view are, roughly, the following: [...] there are a number of processes, of which the generation of ordinals is one, which seem essentially incapable of terminating [...] Thus it is natural to suppose that the terms generated by such a process do not form a class. And, if so, it seems also natural to suppose that any aggregate embracing all the terms generated by one of these processes cannot form a class. Consequently there will be (so to speak) a certain limit of size which no class can reach; and any supposed class which reaches or surpasses this limit is an improper class, i.e. is a non-entity. The existence of self-reproductive processes of this kind seems to make the notion of a totality of all entities an impossible one [...]. (Russell 1906: 43)

---

[20] The term is Potter's (2009: 198).

Thus, according to Russell, the reason why properties having a certain size do not determine a set is that they are indefinitely extensible. It seems that in his view the limitation of size doctrine is ultimately to be explained in terms of the Definiteness Idea. This promises to answer the objection to the limitation of size explanation of the paradoxes we considered in the previous section: whilst the limitation of size explanation of the paradoxes ultimately rests on the fact that supposing certain properties to determine a set leads to paradox, the definite conception need not rest on this explanation. Instead, the suggestion goes, we can *derive* the limitation of size doctrine from the definite conception, similarly to what happened in the case of the iterative conception.

An argument to this effect goes back to Russell (1906) and has more recently been developed by Shapiro and Wright (2006). It is based on the following result:

Theorem 5.9.1. *Assume Property Separation and Property Replacement. Then a property F is definite if and only if* On $\not\preceq$ F.

Proof. Suppose that On $\preceq$ F. Then there is some $H \sqsubseteq F$ such that $H \cong$ On. Since On is not definite, by Property Separation, neither is H. And since H is not definite, by Property Separation again, neither is F.

Suppose, on the other hand, that F is not definite. This means that there is a function f which takes each set of Fs to some $f(A)$ which is F but is not in A. Now use transfinite recursion (which is licensed by Property Replacement) to define the following function on the ordinals:

$$g(\alpha) = f(\{g(\beta): \beta < \alpha\}).$$

By transfinite induction it follows that all the $g(\alpha)$s are F. So we have defined a one-to-one function from the ordinals to the Fs and hence On $\preceq$ F. $\quad\square$

If we endorse the definiteness idea, Theorem 5.9.1 then implies that F determines a set just in case On $\not\preceq$ F, that is that Cantor Limitation of Size holds.

However, Theorem 5.9.1 relies on Property Separation and Property Replacement. Because of this, Potter (2009: 199) has objected to the claim that this argument provides us with 'a route from the definite conception to a kind of limitation-of-size theory'.

Let us consider Potter's concerns about Separation first. He argues that it is doubtful whether Property Separation is justified on the definite conception. The view that it is seems implicit in Russell's own discussion, on the grounds that if F is definite and G is a subproperty of F, then G will be definite too. And Dummett (1991: 317) himself explicitly endorses the view: 'it must be allowed', he writes, 'that every concept defined over a definite totality determines a definite subtotality'. Similar remarks occur in Dummett (1993: 441).

However, Dummett does not offer an argument for thinking that every sub-property of a definite property is itself definite, and, as Shapiro (2003: 80) notes, the view that Property Separation holds on the definite conception is far from unproblematic. For instance, Dummett (1963) himself argued that the property of being an arithmetical truth is indefinitely extensible, which means that the property of being the Gödel code of an arithmetical truth is indefinitely extensible. But every such code is itself a natural number, so if Property Separation holds and the property of being a natural number is definite, we should have that the property of being an arithmetical truth is actually definite. One cannot have that the property of being a natural number is definite and that the property of being an arithmetical truth is not whilst holding on to Property Separation. Something has to go. Dummett, for his part, often expressed doubts about the definiteness of the property of being a natural number. But if one wants to retain the Axiom of Infinity, one is forced to either abandon the Dummettian view that the property of being an arithmetical truth is indefinitely extensible or accept that Property Separation fails after all.

Let us now turn to Property Replacement. Here Potter argues that this is best understood as a limitation of size principle and so its use in establishing that the definite conception leads to limitation of size is circular. Now it is true that Replacement is immediately justified on the limitation of size conception. However, it does not follow from this that it is not justified on the definite conception. Compare: the controversial issue of whether Replacement is justified on the iterative conception is not simply settled by pointing out that it is obviously justified on the limitation of size conception.

To avoid the charge of circularity, it seems that the definite conception theorist needs to argue, without appealing to the limitation of size conception, that if F is definite and there is a functional relation between the Fs and the Gs, then F is definite too. And here Potter's concerns for the case of Property Separation seem to apply too, indeed perhaps even in stronger form: whilst it seems plausible, on the face of it, to hold that if F is definite and G is a subproperty of F, then G is definite too, the analogous claim justifying Property Replacement does not enjoy the same immediate plausibility. Of course, it does if one equates definiteness with not being too big, but in this case Potter's complaint that Replacement is working simply as a limitation of size principle would be justified.

What seems to be needed is a detailed reconstruction of the definite conception from which Property Separation and Property Replacement follow. Such a reconstruction has recently been offered by Linnebo (2018), in a deep and detailed analysis of Dummett's remarks on indefinite extensibility. Working within a second-order intuitionistic logic, Linnebo distinguishes between *intensional* and *extensional* definiteness. Thus, he takes a concept F to be *intensionally definite* (in symbols, $\mathrm{ID}(F)$) just in case

$$\forall x_1, \ldots, x_n(Fx_1, \ldots, x_n \vee \neg Fx_1, \ldots, x_n).$$

And he takes a concept F to be *extensionally definite* (in symbols, $\mathrm{ED}(F)$) if and only if F is intensionally definite and quantification restricted to F preserves intensional definiteness, that is

$$\forall X(\mathrm{ID}(Y) \to \mathrm{ID}((\forall x_1, \ldots, x_n(Fx_1, \ldots, x_n \to Xx_1, \ldots, x_n)))).$$

According to Linnebo's reconstruction of Dummett's arguments, the existence of indefinitely extensible concepts such as that of a set and that of an ordinal shows that there are concepts which are intensionally but not extensionally definite. Linnebo then shows that these definitions of intensional and extensional definiteness intuitionistically imply principles of Separation and Replacement for extensional definiteness, which then immediately deliver Property Separation and Property Replacement.[21]

There are, however, two issues with Linnebo's analysis. The first one is what concepts are sanctioned to be extensionally definite. From Linnebo's definitions and assumptions it would seem to follow that the universal concept is extensionally definite. For he assumes identity to be intensionally definite, and quantification restricted to the universal concept trivially preserves intensional definiteness, since the intersection of any concept F with the universal concept is simply F. This is a problematic result, since the universal concept is usually taken to be a paradigmatic case of a concept which is not (extensionally) definite. Indeed, it seems that Linnebo himself would want the universal concept not to be extensionally definite. In particular, Linnebo does not define a concept to be indefinitely extensible if and only if it is not (extensionally) definite. Instead, he defines a concept F to be indefinitely extensible if and only if the following holds

$$\forall X(\mathrm{ED}(X) \wedge \forall x(Xx \to Fx) \to F(\delta(X)) \wedge \neg X(\delta(X))).$$

This definition closely resembles the definition of indefinite extensibility which, following Russell, we gave in Chapter 1. However, there are some notable differences. First, this definition replaces set-theoretic terminology with second-order

---

[21] Linnebo proves that his definitions also imply further principles. In one case, the proof also makes use of the assumption that identity is extensionally definite. See Linnebo 2018: 214–215.

quantification. Second, Linnebo's definition requires that the concepts over which the second-order variable X ranges should be extensionally definite. This immediately implies that no indefinitely extensible concept is extensionally definite. If we avail ourselves of the resources used to prove Cantor's Theorem, in particular Separation (see Section 1.B), we can then show that the universal concept is indefinitely extensible, which implies that this concept is not extensionally definite.

The second issue with Linnebo's analysis concerns the role of the definition of extensional definiteness in securing Property Separation and Property Replacement. The definition *immediately* delivers a principle of Separation for extensional definiteness. Whilst this is what the defender of the definite conception needed, the legitimate worry in this case is that the definition is just tailored to securing this and other principles needed (such as a principle of Replacement for concept definiteness). This worry is exacerbated by the fact that the definition of extensional definiteness plays no role in the standard arguments showing that a concept is not extensionally definite. For these arguments proceed by showing that the relevant concept is indefinitely extensible which, given the definition of indefinite extensibility, immediately implies that the concept is not extensionally definite. By contrast, if, as is customary, one took a concept to be indefinitely extensible if and only if it is not extensionally definite, no principle of Separation for extensional definiteness (or similar other principles) would follow.

Thus, the issue whether the definite conception implies some sort of limitation of size principle remains open. To be sure, the viability of the definite conception does not depend on whether it entails some limitation of size principle. But it does depend on whether it has the resources to develop some amount of set theory. Pending some further investigation of the definite conception and what it entails, this issue remains open too. In the meantime, note that the iterative conception *explains* the key insight of the definite conception that, when faced with the paradoxes, we ought to retain indefinite extensibility at the expense of universality. Moreover, and more importantly, the definite conception also seems to face issues similar to those faced by the limitation of size conception, namely that properties such as the ordinal property or the universal property are taken not to determine a set because the contrary assumption would lead to paradox.

5.10 Conclusion

In this chapter, we began to explore the idea that the naïve conception is to be modified by restricting the thesis that every property determines a set to certain non-pathological properties. We started by considering the natural suggestion that the naïve conception should be restricted to those properties that do not give rise to inconsistency. We saw that this suggestion either fails to dispense with the inconsistency of naïve set theory or leads to an embarrassment of riches. We then turned to the idea that the naïve conception ought to be restricted to those properties that do not apply to too many things, which gives rise to the limitation of size conception. Whilst this conception can be developed in a variety of ways, its various forms all share the problem that the explanation of the paradoxes they offer has to rest, ultimately, on the fact that supposing certain concepts to be set-forming leads to paradox. This is in contrast with the iterative conception, which explains this fact by reference to the cumulative hierarchy structure of the set-theoretic universe.

The definite conception promises to do better than the limitation of size conception in this respect, since it does not start from some limitation of size principle but has the resources to derive it, or so it has been argued. However, it is doubtful whether the definite conception does have the resources to derive a limitation of size principle. For the usual argument to this effect uses the principles of Property Separation and Property Replacement, and it is not clear whether these principles are justified on the definite conception. It may be that this problem can be overcome: perhaps it is indeed possible to show that Property Separation and Property Replacement are justified on the basis of the definite conception, or perhaps one can develop set theory on the basis of the definite conception in an altogether different manner. However, the apparent advantage of the definite conception over the limitation of size conception turns out to be illusory: the definite conception too, in the end, has to explain the paradoxes on the basis of the fact that supposing certain concepts to determine a set leads to paradox.

Appendix 5.A Generalizing McGee's Theorem

We state and prove the generalization of McGee's Theorem due to myself and Murzi. We first need some terminology and definitions. If T is a theory, we write '$\Gamma \vdash_T \varphi$' for '$\Gamma \cup T \vdash \varphi$'. Then we make use of the following definitions:

DEFINITION 5.A.1. Let Γ be a set of sentences and Σ a first-order schema.

(i) Γ is T-*consistent* iff $\Gamma \nvdash_T \bot$.
(ii) Γ is Σ-*maximally T-consistent* iff Γ is T-consistent and any set which properly extends Γ and contains Σ-instances not in Γ is T-inconsistent.
(iii) Γ is *maximally T-consistent* iff it is T-consistent and any set which properly extends Γ is T-inconsistent.

Now let S be a consistent first-order theory entailing the axioms of some theory S', let Σ be a first-order schema, and let $\mathcal{L}$ be a countable first-order language. Suppose we relatively interpret in $\mathcal{L}$ the language obtained by adding to the language of S any non-logical expression in Σ. We then have:

Theorem 5.A.2 (Incurvati-Murzi). *Suppose that for each φ in $\mathcal{L}$ there is an instance Σ_φ of Σ such that*

$$\vdash_{S'} \varphi \leftrightarrow \Sigma_\varphi.$$

Then:

(1) *For any S-consistent set Δ of sentences in $\mathcal{L}$, there is a Σ-maximally S-consistent set Γ of Σ-instances such that $\Gamma \vdash_S \delta$ for every $\delta \in \Delta$.*

(2) *If Ξ is a Σ-maximally S-consistent set of Σ-instances, then, for every ψ in $\mathcal{L}$, $\Xi \vdash_{S'} \psi$ or $\Xi \vdash_{S'} \neg\psi$.*

Proof. Enumerate the sentences of $\mathcal{L}$ as $\psi_0, \psi_1, \psi_2, \dots$ and define an increasing sequence of finite sets of Σ-instances as follows:

$$\Gamma_0 = \{\Sigma_\delta : \delta \in \Delta\};$$

$$\Gamma_{n+1} = \begin{cases} \Gamma_n \cup \{\Sigma_{\psi_n}\} & \text{if } \Gamma_n \cup \{\psi_n\} \text{ is S-consistent;} \\ \\ \Gamma_n \cup \{\Sigma_{\neg\psi_n}\} & \text{otherwise.} \end{cases}$$

Now let $\Gamma = \bigcup_{n \geq 0} \Gamma_n$. Clearly, $\Gamma \vdash_S \delta$ for every $\delta \in \Delta$, since $\Gamma \supseteq \Gamma_0$ and

$$\vdash_S \delta \leftrightarrow \Sigma_\delta$$

for every $\delta \in \Delta$. Moreover, Γ_0 is S-consistent if Δ is, and the sequence $\Gamma_0, \Gamma_1, \Gamma_2, \dots$ is defined so that if Γ_n is S-consistent, then Γ_{n+1} is S-consistent too. By induction on n, it follows that if Δ is S-consistent, then so is Γ_n for each n, and hence Γ itself.

It remains to check that Γ is a Σ-maximally S-consistent set of Σ-instances. So take a ψ in $\mathcal{L}$ and suppose $\Sigma_\psi \notin \Gamma$. ψ must appear somewhere in the list $\psi_0, \psi_1, \psi_2, \dots$, that is $\psi = \psi_m$ for some m. But if $\Sigma_{\psi_m} \notin \Gamma$, this must be because $\Gamma_m \cup \{\psi_m\}$ is S-inconsistent. And this means that $\Gamma_m \cup \{\Sigma_{\psi_m}\}$ is S-inconsistent, since

$$\vdash_S \psi_m \leftrightarrow \Sigma_{\psi_m}.$$

Since $\Gamma_m \subseteq \Gamma$, $\Gamma \cup \{\Sigma_\psi\}$ is S-inconsistent too.

For (2), let Ξ be a Σ-maximally S-consistent set of Σ-instances, and take a ψ in $\mathcal{L}$. Since Ξ is S-consistent, then either $\Xi \cup \{\psi\}$ or $\Xi \cup \{\neg\psi\}$ is S-consistent. So either $\Xi \cup \{\Sigma_\psi\}$ or $\Xi \cup \{\Sigma_{\neg\psi}\}$ is S-consistent. But since Ξ is Σ-maximally S-consistent, either $\Sigma_\psi \in \Xi$ or $\Sigma_{\neg\psi} \in \Xi$. Thus, $\Xi \vdash_{S'} \psi$ or $\Xi \vdash_{S'} \neg\psi$. $\square$

We can now apply the above generalization of McGee's Theorem to obtain the desired result for the case of set theory and the Naïve Comprehension Schema. To this end, we use Boolos's observation mentioned in the chapter (i.e. Lemma 5.1.1). Let R be some consistent theory entailing the axioms of the theory whose sole non-logical axiom is (Comp$_\emptyset$), i.e. the instance of (Comp) asserting the existence of the empty set. And suppose that we have relatively interpreted the language of R in a countable first-order language $\mathcal{L}'$. We have:

Lemma 5.A.3. *For each φ in $\mathcal{L}'$,*

$$\vdash_{Comp_\emptyset} \varphi \leftrightarrow \exists y \forall x (x \in y \leftrightarrow (\neg\varphi \ \& \ x \notin x)).$$

Proof. Immediate from Lemma 5.1.1. $\square$

But together with Theorem 5.A.2, Lemma 5.A.3 immediately yields:

Corollary 5.A.4.

(1) For any R-consistent set Δ of sentences in $\mathcal{L}'$, there is a (Comp)-maximally R-consistent set Γ of (Comp)-instances such that $\Gamma \vdash_R \delta$ for every $\delta \in \Delta$.

(2) If Ξ is a (Comp)-maximally R-consistent set of (Comp)-instances, then, for every ψ in $\mathcal{L}'$, $\Xi \vdash_{Comp_\emptyset} \psi$ or $\Xi \vdash_{Comp_\emptyset} \neg\psi$.

6

The Stratified Conception

The systems presented by W. V. Quine
mediated, as it were, between type theory
and the systems of axiomatic set theory.

Paul Bernays (1971: 172)

In Chapter 5, we focused on the attempt to restrict the naïve conception to properties that do not apply to too many things. This is a *semantic* restriction: whether a property determines a set does not depend on the way it is expressed, but rather on the size of its extension.

In this chapter, we investigate *syntactic restrictionism* – the idea that the Naïve Comprehension Schema is to be restricted according to *syntactic* notions. The syntactic notion we shall focus on is that of *stratification*. Restricting the Naïve Comprehension Schema to those formulae that are stratified gives rise to a set theory due to Quine. As we shall see, this theory can be seen as embodying what I call the *stratified conception of set*. I argue, however, that this conception is best regarded as a conception of *objectified property*. This allows one to overcome objections that can be raised against the conception as a conception of *set*. Taking the stratified conception to be a conception of objectified property also enables one to further develop the diagnosis of the paradoxes presented in Chapter 1. On this extended diagnosis, the paradoxes arise from conflating the concept of set and that of objectified property.

6.1 The Early History of Syntactic Restrictionism

According to syntactic restrictionism, a property determines a set just in case its syntactic expression is of a certain kind. The idea goes back at least to Russell

(1906: 29) who, as mentioned in the previous chapter, considered three strategies for dealing with the paradoxes. One of these strategies, recall, is pursued by the *zigzag theory*, which embodies a form of syntactic restrictionism. Russell introduces it as follows:

In the zigzag theory, we start from the suggestion that propositional functions determine classes when they are fairly simple, and only fail to do so when they are complicate and recondite. If this is the case, it cannot be bigness that makes a class go wrong; for such propositional functions as 'x is not a man' have an exemplary simplicity, and are yet satisfied by all but a finite number of entities. (Russell 1906: 38)

As Russell goes on to explain, the name 'zigzag theory' is due to the fact that, if φ is a formula which does not determine a set, then, for every set a, there must be either members of a to which φ does not apply, or members of a's complement $\bar{a}$ to which φ does apply. For otherwise we would have that φ applies to all and only members of a, contradicting the fact that it does not determine a set. A formal representation of the situation will make the zigzag nature of φ more apparent:

$$\exists y(y \in a \land \neg\varphi(y))$$
$$\lor$$
$$\exists y(\varphi(y) \land y \in \bar{a}).$$

Russell points out a number of features of the zigzag theory, which markedly distinguish it from theories based on the iterative conception or limitation of size. To start with, Russell claims, if φ is simple, so is its negation. It follows that each set has a complement (including the empty set, whose complement is the universal set). In iterative set theory, *no* set has a complement.[1]

Moreover, the zigzag theory follows the Frege-Russell definition of cardinals as equivalence classes of equinumerous sets (see Appendix 1.A). This definition is not available in iterative set theory, since for any set other than the empty set, the set of sets equinumerous with it is not bounded in the hierarchy. Similarly, in the zigzag theory one can define ordinals as equivalence classes of well-ordered sets. Again, this definition is not available in iterative set theory, for essentially the same reason.

Finally, the zigzag theory promises a different treatment of the paradoxes from iterative set theory. Russell tells us that the zigzag theory asserts the existence of the greatest cardinal but avoids Cantor's Paradox by denying that the instance of Comprehension used in the proof of Cantor's Theorem is simple. In the case of ordinals, the zigzag theory holds that there is a set of all of them but there is no greatest ordinal: '[a]ll that Burali-Forti's contradiction forces us to admit is that

---

[1] For suppose some set a did have a complement. Then, $a \cup \bar{a}$ would exist. But this set is just the universal set.

there is no *maximum* ordinal, i.e. that the function "α and β are ordinal numbers, and α is less than β" and all other functions ordinally similar to this one are non-predicative [i.e. not simple]' (1906: 39). Perhaps more perspicuously, what Russell seems to have in mind here is that although there is a set of all ordinals and this set has a well-ordering Ω, there is no ordinal given by the set $\{\alpha : \alpha < \Omega\}$ since, in general, the condition $\alpha < \beta$ is not simple.

Although he gives examples of simple formulae, Russell (1906) does not give axioms for simplicity. He had tried to do so in 1904, when working on various versions of the zigzag theory. He never succeeded – or better, he was never satisfied with the result: he believed his proposed axioms to be complicated and lacking in motivation. He took this lack of motivation to be especially problematic, exposing the theory to the risk of contradictions. Here, it is perhaps worth remarking that some of the versions of the zigzag theory Russell developed turned out to be inconsistent. Eventually, Russell embraced the no class theory – the third of the strategies considered in Russell 1906 – only to come full circle and return to the theory of types (see Urquhart 1988).

However, some thirty years later, a theory was developed that bears striking similarities to the zigzag theory but whose axioms are less cumbersome than those formulated by Russell. It is to this theory that we now turn.

6.2 New Foundations and Cognate Systems

In his 'New foundations for mathematical logic' Quine (1937) presents a form of syntactic restrictionism based on the idea that the Comprehension Schema should be restricted to its *stratified* instances. We say that a formula φ in the language of set theory is *stratified* just in case it is possible to assign natural numbers to its variables in such a way that (i) for any subformula of φ of the form $x = y$ the same natural number is assigned to x and y and (ii) for any subformula of φ of the form $x \in y$, y is assigned the successor of the number assigned to x.[2] New Foundations (NF) is then the theory consisting of Extensionality and the *Stratified Comprehension Schema*, the axiom schema whose instances are all stratified instances of (Comp). That is to say, Stratified Comprehension consists of all instances of

$$\exists y \forall x (x \in y \leftrightarrow \varphi(x)), \tag{SComp}$$

---

[2] Quine (1937) only stated condition (ii) in his definition of a stratified formula. Condition (i) follows in his system because identity is defined in terms of membership by stipulating that $x = y$ is to mean that for any set z, x belongs to z iff y does.

where $\varphi(x)$ is any *stratified* formula in $\mathcal{L}_\in$ in which x is free and which contains no free occurrences of y.[3]

The standard reasoning leading to Russell's Paradox is blocked in NF: since the formula $x \notin x$ is unstratified, we do not have the instance of Comprehension asserting the existence of the Russell set. Mirimanoff's Paradox is dealt with in a similar manner: since the formula saying that x is well-founded is unstratified, we do not have the instance of Comprehension asserting the existence of the set of all well-founded sets. To this extent, NF follows iterative set theories such as ZF. Unlike ZF, however, NF asserts the existence of the universal set V, since the formula $x = x$ is obviously stratified. The standard proof of the paradox of the set of all sets is blocked because the condition used to define the function that should allow us to diagonalize out of V, namely $f(x) = \{y \in x : y \notin y\}$, is not stratified. For the same reason, in NF one cannot simply deduce the existence of the Russell set from the existence of the universal set: the relevant instance of ZF's Separation Schema corresponds to an unstratified instance of Naïve Comprehension.[4]

Not only does NF admit the existence of V; in NF, every set has a complement. Indeed, the universe of NF gives rise to a Boolean algebra with the operations of union, intersection and complementation. Other operations disallowed in iterative set theory but available in NF are the operation of forming the set of all sets of which a set is a member and the operation of forming the set of all sets of which a set is a subset. That is, given any set a, the sets $\{x : a \in x\}$ and $\{x : a \subseteq x\}$ always exist.

In NF, we cannot use the von Neumann definition of an ordinal (see Section 1.A), since we cannot prove that every well-ordered set is isomorphic to a von Neumann ordinal. Nor can we use the Scott-Tarski trick (see Section 3.2), since that makes use of the Axiom of Foundation.[5] However, in NF one can simply use the Frege-Russell definition of a cardinal as an equivalence class of equinumerous sets, and the definition of an ordinal as an equivalence class of well-orderings under isomorphism. And once these definitions are adopted, it is straightforward to check that Stratified Comprehension implies the existence of the set of all cardinals and the set of all ordinals.

---

[3] Notably, although Stratified Comprehension has infinitely many instances, these instances can be deduced from finitely many axioms, which can be in turn deduced from instances of (SComp). Thus, NF is finitely axiomatizable. The finite axiomatizability of NF was established by Hailperin (1944). Simpler axiomatizations than Hailperin's were subsequently devised by Crabbé and Holmes among others.

[4] Similarly, one cannot derive the existence of the set of well-founded sets from the existence of V, since the relevant instance of Separation corresponds to an unstratified instance of Naïve Comprehension.

[5] As noted in Section 3.2, the Scott-Tarski trick actually requires only Coret's Axiom B, which is provable in ZFC minus the Axiom of Foundation (and hence in the theories to be discussed in the next chapter). Coret's Axiom, however, does not hold in NF.

So what about Cantor's Paradox of the set of all cardinals? Recall that this paradox is obtained by considering a recipe which, for any cardinal κ, produces a cardinal larger than κ. Given the existence of a set of all cardinals, we obtain a contradiction. In iterative set theory, the recipe is provided by Cantor's Theorem, which must therefore fail in NF, on pain of inconsistency. And indeed, the standard proof is blocked, since it makes use of an unstratified formula.[6] Nonetheless, as Quine (1938a) pointed out, one can prove in NF that $\mathcal{P}(a)$ is larger than $\mathcal{P}_1(a)$, the set of all singleton subsets of a.[7] Following Rosser (1953), say that a set a is *cantorian* if it is equinumerous with $\mathcal{P}_1(a)$. Cantor's Paradox becomes a proof that V is not cantorian. Cantor's Theorem becomes a proof that if a set is cantorian, it is smaller than its powerset. Indeed, this is an instance of a general phenomenon in NF: many theorems of iterative set theory fail in general but hold when relativized to cantorian sets.

As for the Burali-Forti Paradox, recall that this is obtained by considering a recipe which, for any segment of the ordinals, produces an ordinal not in that segment. In particular, for any α, we consider the ordinal number of the series of ordinals less than α under the well-ordering relation $<$. In iterative set theory, we can then prove by transfinite induction that this ordinal number must be α itself. This gives rise to a contradiction when α is the ordinal number Ω of the set of all ordinals. However, the proof that the ordinal number of the series of the ordinals up to α is α makes use of an unstratified instance of Comprehension and is therefore blocked in NF (see, e.g., Rosser 1942: 2–3). Once again, NF asserts the existence of a certain big set, and paradox is avoided because we cannot diagonalize out of it.

Now let us pause on the fact that NF proves that the series of ordinals less than Ω is well-ordered by $<$ and hence has an ordinal number, Ω_0. On pain of contradiction, $\Omega_0 \neq \Omega$ and so $\Omega_0 < \Omega$. But now consider the series of ordinals less than Ω_0. This too is well-ordered by $<$ and has an ordinal number, Ω_1. As before, we must then have that $\Omega_1 < \Omega_0$. From an external point of view, we have what appears to be an infinite descending sequence of ordinals. However, this does not contradict NF's theorem that the set of all ordinals is well-ordered by $<$; it just shows that, if consistent, NF must think that the sequence $\Omega > \Omega_0 > \Omega_1 > \ldots$ is not a set. This is sometimes put by saying that NF has no standard model.[8]

---

[6] Cantor asks us to consider the set $b = \{x \in a : x \notin f(x)\}$, but as Quine already observed in his original paper, the condition $x \notin f(x)$ is not stratified.

[7] In particular, one can simply adapt Cantor's original proof and consider the set $b = \{x \in a : \neg(\{x\} \subseteq f(\{x\}))\}$. Since the condition $\neg(\{x\} \subseteq f(\{x\}))$ *is* stratified, the proof goes through in NF.

[8] Rosser and Wang (1950) proved that there is no model of NF in which identity is standard, the finite cardinals are well-ordered, and the ordinals are well-ordered. It was later shown that NF proves that every finite cardinal can be well-ordered.

What is the strength of NF compared to that of ZFC? A first possible answer here is to consider which ZFC axioms are provable in NF. Besides Extensionality, which is an axiom, NF proves the Axioms of Unordered Pairs, Union and Powerset (Hailperin 1944). It refutes Separation and Replacement but proves their stratified versions. It refutes Foundation, since it proves that $V \in V$. What about Infinity and Choice? Specker (1953a) proved that AC fails in NF and derived Infinity as a corollary.[9] Specker's proof is technical, but the conclusion that Choice fails in NF can be made plausible as follows. For any cardinal κ, let $T(\kappa)$ be the cardinal of $\mathcal{P}_1(a)$ when the cardinal of a is κ. And let ∞ be the cardinality of the universal set. Then, similarly to the case of ordinals, there is an infinite descending sequence $\infty > T(\infty) > T(T(\infty)) > \dots$. Thus, the cardinals are not well-ordered, but in this case (unlike the case of ordinals), this can be shown *internally* (that is, in NF). But since Zermelo's proof that AC implies that all sets are well-ordered can be reconstructed in NF, it follows that the Axiom of Choice must fail. This also yields the Axiom of Infinity, since one can show in NF that all finite cardinals are well-ordered.[10] Note that Specker's result is compatible with Choice holding for cantorian sets: pending an argument to the effect that V ought to be well-orderable, this may go some way towards mitigating worries about the failure of AC.[11]

But in asking about the strength of NF compared to that of ZFC, one might be wondering about its *consistency* strength with respect to ZFC.[12] This question relates, more generally, to the consistency problem for NF, that is the problem of trying to prove its consistency in ZFC (possibly extended with some large cardinal axiom). Doubts about the consistency of NF arose when Rosser (1942) (and independently Lyndon) showed that the Burali-Forti paradox is derivable in ML, an extension of NF with classes adopted by Quine (1940). However, the inconsistency of ML does not immediately transpose to that of NF, and a version of ML which is equiconsistent with NF was developed by Wang (1950). At the time of writing, the consistency problem for NF is still open, but a manuscript by Randall Holmes is under review in which a purported consistency proof for NF is given.[13] If correct,

---

[9] Quine (1937) had claimed that NF proves Infinity, but his argument was erroneous. For it was based on the fact that NF proves the existence of infinitely many objects, which falls short of establishing that these objects form a set and hence that V is infinite.

[10] In his paper, Specker mentions another corollary of his proof, namely that the Generalized Continuum Hypothesis – the assertion that for no infinite cardinal κ is there a cardinal μ such that $\kappa < \mu < 2^\kappa$ – fails, since the proof that it implies AC can be formalized in NF.

[11] See Section 6.9 for more on the status of AC in NF.

[12] A theory T has greater consistency strength than theory S if the consistency of T implies the consistency of S but not vice versa.

[13] Murdoch Gabbay also has a manuscript (see https://arxiv.org/abs/1406.4060) in which a purported proof of the consistency of NF is given, using different techniques.

the proof shows that the consistency strength of NF is rather weak, being the same as that of MacLane set theory (MacLane 1986), that is Z with Separation restricted to bounded formulae.[14]

Jensen (1969) considered the theory NFU consisting of Stratified Comprehension and *Weak Extensionality*, that is Extensionality restricted to objects having elements, i.e. non-*Urelemente* other than the empty set. He proved the consistency of NFU relative to that of PA and the consistency of NFU plus AC and Infinity relative to that of Z. The consistency strength of the latter theory is now known to be exactly that of MacLane set theory. More recently, Crabbé (1992) has shown that the theory consisting of Stratified Comprehension alone interprets NFU: in the context of Stratified Comprehension, Weak Extensionality does not increase consistency strength.

Finally, what about the mathematics that can be carried out in NF? Rosser (1953) showed how to develop the basics of analysis and set theory in NF. As Rosser observes, one can deal with anomalies arising from the failure of AC by assuming that Choice holds for cantorian sets. An alternative course of action, pursued by Holmes (1998), is to use NFU and add Choice (as well as Infinity) as an axiom.

In the course of his development of mathematics within NF, Rosser noticed some difficulties arising from the unavailability of induction on unstratified formulae. To repair the situation, he adopted an *Axiom of Counting*, which asserts that for each natural number n, the set of all numbers less than or equal to n is in n. The Axiom of Counting is weaker than full induction on stratified and unstratified conditions alike, but increases the consistency strength of the theory, as later showed by Orey (1964). Orey's result actually concerned a statement equivalent to Counting, namely the assertion that the set of natural numbers is strongly cantorian, where a set a is *strongly cantorian* if there is a function that sends each member of a to its singleton. In NF, assertions about certain sets being cantorian or strongly cantorian have an effect similar to that of axioms of infinity in iterative set theory. For instance, Henson (1973) considered an axiom stating that every cantorian set is strongly cantorian, which implies the Axiom of Counting and whose addition to NFU + Infinity yields a theory equiconsistent with ZFC + 'There is an n-Mahlo cardinal' for each n.[15] Stronger axioms in the same spirit were stated and studied by Holmes (1998; 2001).[16]

---

[14] A formula φ of $\mathcal{L}_\in$ is *bounded* iff all quantifiers in φ are restricted to $x \in t$ for some term t.

[15] A cardinal κ is α-*Mahlo* if k is inaccessible and for every $\beta < \alpha$, the set of β-Mahlo cardinals below κ is stationary in κ.

[16] For more on NF and, more generally, set theories which admit of a universal set, see Forster 1995.

6.3 Rejecting Indefinite Extensibility

Is NF a zigzag theory? At first sight, it might seem that it cannot be. For even if φ is not stratified, there can be a set of all and only the things to which φ applies. Consider the following example (see Fraenkel et al. 1973: 162–163, fn. 2). Take the set $z \cup \{z\}$. The instance of (Comp) asserting the existence of this set is unstratified. But using the instance of (SComp) corresponding to the stratified formula $x \in z \vee x = u$, one can prove that for all z and u, $z \cup u$ exists, and hence that $z \cup \{z\}$ exists.

However, suppose that we relax the stratification requirement in Stratified Comprehension so as to apply only to the *bound* variables in $\varphi(x)$ (as well as x). The resulting axiom schema – *Weakly Stratified Comprehension* – entails the existence of $z \cup \{z\}$ without any *detour* through the instance of Stratified Comprehension formed using $x \in z \vee x = u$. And the result of replacing Stratified Comprehension with Weakly Stratified Comprehension gives rise to the same theory as NF. Thus, if a formula is simple if and only if it is weakly stratified, then NF might qualify as a zigzag theory.

Indeed, the similarities between NF and the zigzag theory are striking. In both theories, every set has a complement. Both theories assert the existence of big sets such as the universal set, the set of all cardinals and the set of all ordinals. And, finally, both theories block the paradoxes not by denying the existence of these sets, but by rejecting the instance of Naïve Comprehension used to diagonalize out of them.

However, the similarities between the two theories should not be exaggerated. For one thing, the details of the zigzag theory as presented by Russell are too vague to properly assess the matter. For another, some of the details that Russell does provide point to differences between the two theories. For instance, Russell tells us that the zigzag theory deals with the Burali-Forti paradox by denying that $\{\alpha : \alpha < \Omega\}$ is a set. By contrast, in NF there is such a set but paradox is avoided because its ordinal number is not Ω.

This difference in the treatment of the Burali-Forti Paradox notwithstanding, there seems to be a common strategy adopted to deal with the paradoxes by the zigzag theory and NF (as well as cognate systems such as NFU). To see this, recall that, according to the diagnosis offered in Section 1.7, the paradoxes arise because on the naïve conception we have both indefinite extensibility and universality of the set concept. And recall, further, that the iterative conception is an attempt to articulate a conception of set which motivates indefinite extensibility and the rejection of universality. But, one might ask, why do we have to proceed in this way? Why can't we try to articulate a conception of set which motivates universality and the rejection of indefinite extensibility? If our diagnosis of the paradoxes is correct, this should be an equally acceptable way of dealing with them from a mathematical point of view.

There are some limitations to this way of proceeding, however. For on Russell's formalization, indefinite extensibility looked as follows:

$$\forall u (\forall x (x \in u \rightarrow \varphi(x)) \rightarrow (f(u) \notin u \wedge \varphi(f(u)))). \tag{IE}$$

In the context of this formalization, then, the suggestion to reject indefinite extensibility amounts to saying that when φ and f satisfy (IE), one should deny the existence of $f(w)$, where w is the set of all φs. And Russell's formalization makes it clear that this strategy does not seem available in the case of Russell's Paradox, where we let $\varphi(x)$ be $x \notin x$ and $f(x) = x$: it does not seem an option to deny that there is a function that sends every object to itself. And indeed, both the zigzag theory and NF follow iterative set theory here and reject the existence of the Russell set.

But it might be an option to deny the existence of f for different choices of $\varphi(x)$ and f. Consider the Burali-Forti Paradox, for instance. Given (IE), this is obtained by taking $\varphi(x)$ to be 'x is an ordinal' and f to be the function which sends a set a to the least ordinal number bigger than all sets in a. It is not immediately obvious that in this case we should deny that there is a set of all φs rather than that $f(x)$ should exist for any argument x. Similar considerations apply to Cantor's Paradox. Both the zigzag theory and NF pursue this line of thought: although they deny the existence of the Russell set, they allow for the existence of the set of all sets, the set of all ordinals and the set of all cardinals by denying that, for the relevant f, $f(x)$ exists for all xs.

This option was already suggested by Russell. He writes:

The theory of limitation of size neglects the second alternative (that $f(w)$ may not exist), and decides for the first (that φ is not predicative). Thus, in the case of the series of ordinals, the second alternative is that the whole series of ordinals has no ordinal number, which is equivalent to denying the predicativeness of 'α and β are ordinal numbers, and α is less than β'. The adoption of this alternative would enable us to hold that all ordinals do form a class, and yet there is no greatest ordinal. But the theory in question rejects this alternative, and decides that the ordinals do not form a class. The only case in which this is the only alternative is when $f(u)$ is u itself; otherwise we always have a choice. (Russell 1906: 44, with notation slightly changed)

Indeed, since Russell himself took the zigzag theory to deny the predicativeness of 'α and β are ordinal numbers, and α is less than β', it seems clear that he took the zigzag theory as embracing the second alternative and hence as rejecting indefinite extensibility and retaining universality. What we see is that the same holds for NF, setting aside the differences in the treatment of the Burali-Forti Paradox. In general, as we saw in the previous section, in NF the arguments used to diagonalize out of big sets such as the universal set, the set of all cardinals and the set of all ordinals are blocked. And diagonal arguments are what is typically used to show that a certain concept displays indefinite extensibility.

6.4 The Received View on NF

NF is a remarkably simple system, within which one can reconstruct a substantial amount of ordinary mathematics. Yet it is widely considered unsuitable as a foundational theory.

This point of view goes back at least to Russell (1959). According to him, a satisfactory solution to the paradoxes should satisfy three conditions. First, it should be free of contradictions. Second, it should preserve as much mathematics as possible. And third, it should preserve what Russell called 'logical common sense', i.e. once the proposed solution is put forward, it should seem that the solution is what one was to expect all along. Russell went on to say that the third condition

is not regarded as essential by those who are content with logical dexterity. Professor Quine, for example, has produced systems which I admire greatly on account of their skill, but which I cannot feel to be satisfactory because they seem to be created *ad hoc* and not to be such as even the cleverest logician would have thought of if he had not known of the contradictions. (Russell 1959: 61)

Russell's verdict on NF is in line with his earlier assessment of the zigzag theory: although the axioms of NF are less complicated than those of the zigzag theory, they still lack intrinsic motivation.

With the emergence of the iterative conception as the dominant paradigm in the early 70s, the charge of ad hocery against NF became common currency.

New Foundations is not the axiomatization of an intuitive concept. It is the result of a purely formal trick intended to block the paradoxes. No further axioms are suggested by this trick. Since there is no intuitive concept, one is forced to think in terms of the formal axioms. Consequently, there has been little success in developing New Foundations as a theory. (Martin 1970: 113)

[Theories like NF and ML] appear to lack a motivation that is independent of the paradoxes [. . .] A final and satisfying resolution to the set-theoretical paradoxes cannot be embodied in a theory that blocks their derivation by artificial technical restrictions on the set of axioms that are imposed *only because* paradox would otherwise ensue. (Boolos 1971: 17)

From the point of view of the philosophy and the foundations of mathematics the main drawback of NF is that its axiom of comprehension is justified mostly on the technical ground that it excludes the antinomic instances of the general axiom of comprehension, but there is no mental image of set theory which leads to this axiom and lends it credibility. (Fraenkel et al. 1973: 164)

Notably, Quine (1970) replied to Martin, thus breaking his self-reported habit of not responding to reviews. Even more notably, he only challenged some of the technical assertions Martin had made in his scathing review. He did not question Martin's main point that there is no conception of set behind NF. Arguably, this is because Quine did not take this point to be particularly worrisome. As we saw in the previous chapter, Quine believed that the only really intuitive conception of set is the naïve conception. Given the inconsistency of the theory to which the naïve conception gives rise, the best we can hope for is a theory which blocks the paradoxes whilst retaining a substantial amount of mathematics.

    According to Quine, then, there is no intuitive conception of set behind any set theory other than the naïve one. And on top of this, Zermelo-style systems are hopelessly artificial, whilst NF naturally emerges from the theory of types, as we shall see in the next section: the set theory guilty of ad hocery is actually ZF. What Quine had apparently overlooked is that Zermelo-style theories too can be naturally obtained by lifting certain restrictions imposed by type theory:

in its multiplicity of axioms [Zermelo's system] seemed inelegant, artificial, and ad hoc. I had not yet appreciated how naturally his system emerges from the theory of types when we render the types cumulative and describe them by means of general variables. (Quine 1987a: 287)

Once he appreciated the connection between type theory and iterative set theory, Quine went on to detail it in print (Quine 1956). This connection had in fact already been described by Gödel (*1933). In the next section, we shall see how both the cumulative hierarchy and NF can be seen as being obtained by removing certain restrictions of type theory. This will pave the way towards the stratified conception.

6.5 From Type Theory to NF

Let us start by describing the basics of the theory of types, first introduced by Russell (1903: Appendix B) and subsequently developed by Whitehead and Russell (1910, 1912, 1913). So far we have considered theories cast in first-order logic, whose variables range over objects, and theories cast in second-order logic, which also have second-order variables, which can be taken as ranging over concepts, or properties, or pluralities. But one could consider adding an infinite hierarchy of variables and quantifiers, so as to obtain a hierarchy of languages of order n for each n. The result is the (simple) *theory of types*. In this theory, each variable is assigned a natural number, its *type*. On the classical interpretation, variables of type 0 range over individuals, objects which are not classes. Variables of type 1 range over classes of individuals, variables of type 2 range over classes of classes of individuals, and so on. Finally, we impose the condition that expressions that violate the type restrictions are meaningless. Thus, an expression of the form $x \in y$ is meaningful only if the type of y is the successor of the type of x, and an expression of the form $x = y$ is meaningful only if x and y have the same type.

Now what Gödel observed in 1933 is that there is a natural mathematical road from the theory of types to iterative set theory:

[O]nly one solution [to the paradoxes] has been found, although more then 30 years have elapsed since the discovery of the paradoxes. This solution consists in the theory of types. [...] It may seem as if another solution were afforded by the system of axioms for the theory of aggregates, as presented by Zermelo, Fraenkel and von Neumann; but it turns out that this system is nothing else but a natural generalization of the theory of types, or rather, it is what becomes of the theory of types if certain superfluous restrictions are removed. (Gödel *1933: 45–46)

Gödel lists three superfluous restrictions. The first is that the theory of types as formulated by Whitehead and Russell (1910, 1912, 1913) only encompasses the *finite* types. The second is that the types are *disjoint*: no entity can belong to more than one type. The third is that predications that violate the type restrictions are regarded as meaningless. Gödel suggests that iterative set theory can be obtained by lifting these three restrictions. First, we admit infinite types. Second, we allow the types to be cumulative – that is, we allow each type to contain all entities appearing at lower types. And third, we regard predications that violate the type restrictions not as meaningless but simply false. The result is, of course, the cumulative hierarchy. Gödel does not offer a mathematical proof of his claim, but this has been given by Linnebo and Rayo (2012).

There are technical advantages to making the types cumulative. On the theory of types, we have what appear to be, from a type-free point of view, unnecessary repetitions. Consider the set-theoretic interpretation of type theory. Rather than having one empty set, we have one for each type greater than 0: the type 1 set with no members, the type 2 set with no members, and so on. Using the Frege-Russell definition of cardinal number, we have a different number 0 for each type, a different number 1 for each type, and so on. Allowing mixing of types removes such repetitions:

The truth is that there is only one satisfactory way of avoiding the paradoxes: namely, the use of some form of the *theory of types*. That was at the basis of both Russell's and Zermelo's intuitions. Indeed the best way to regard Zermelo's theory is as a simplification and extension of Russell's. [...] The simplification was to make the types *cumulative*. Thus mixing of types is easier and annoying repetitions are avoided. (Scott 1974: 208)

To make life with the 'unnecessary repetitions' easier, Russell had adopted a policy of *typical ambiguity*. According to this policy, we can leave the type of some variables indefinite if it is *possible* to assign a type to each of them so as to obtain a meaningful formula of type theory. In other words, the policy of typical ambiguity tells us that formulae that are *stratified* can be safely used. Note that this does not mean that formulae with unadorned variables are now meaningful. But it means, for instance, that although we officially have different natural numbers for each type, an argument using typical ambiguity might provide a template which delivers, for each n, a proof about the natural numbers occurring at type n.

   According to Quine, adopting a policy of typical ambiguity does not go to the heart of the matter, and he cites the unnecessary repetitions as his motivation for NF:

Not only are all these cleavages and reduplications intuitively repugnant, but they call continually for more or less elaborate technical maneuvers by way of restoring severed connections. (Quine 1937: 79)

To avoid the paradoxes, says Quine, one does not need to ban unstratified formulae *in every context*. One only needs to do so when forming instances of Comprehension. The resulting theory has more deductive power than the theory of types and does not suffer from the anomalies engendered by the type restrictions. By anti-exceptionalism about set theory, Quine seems to suggest, this theory ought to be preferred to the theory of types.

   In a subsequent paper (Quine 1938b), Quine distinguished between two aspects of the theory of types, the *ontological doctrine* and the *formal restriction* (see also Morris 2017). The ontological doctrine consists in regarding the universe as stratified: it consists of a layer of individuals, a layer of classes, a layer of classes of classes, and so on. The formal restriction consists in regarding unstratified formulae as meaningless. Now the formal restriction is often formulated with reference to types and hence to the ontological doctrine. But it does not have to be. And once it is formulated in purely syntactical terms (see Quine 1938b for details), the ontological doctrine is not forced upon us, or so Quine argues. Indeed, rejecting the ontological doctrine suffices to dispense with the intuitively repugnant repetitions:

The type ontology was at best only a graphic representation or metaphysical rationalization of the formal restrictions; and though some such rationalization may well be desired, it seems clear in particular that the type ontology afforded less help than hindrance. (Quine 1938b: 133)

Note that Quine's distinction also sheds light on what one is doing when obtaining the cumulative hierarchy by lifting some restrictions imposed by type theory. In that case, one is rejecting the formal restrictions whilst retaining the ontological doctrine, albeit in modified form: the universe *is* layered, but the layers are cumulative. Our discussion shows that what Scott said of iterative set theory extends to NF: both theories can be traced back to the theory of types.

The foregoing considerations suggests regarding the universe of NF as what is obtained by *collapsing* the type-theoretic hierarchy. This was later confirmed mathematically by Specker (1953b; 1962). Let φ^+ be the result of raising the types of all variables in φ by one. Then $\varphi^+ \leftrightarrow \varphi$ is an axiom of typical ambiguity, and Specker proved that the theory obtained by adding this axiom to type theory is equiconsistent with NF.

Collapsing the types is certainly mathematically natural, as Quine stressed. But is it also conceptually natural? What are we doing, from a conceptual point of view, when we collapse the types? It is to these questions that we now turn.[17]

6.6 The Stratified Conception

We saw that on the classical interpretation, type theory deals with individuals, classes, classes of classes, and so on. But nowadays type theory is mostly regarded as dealing with what are taken to be genuinely higher-order entities, such as concepts, properties or pluralities, which are distinguished from individual objects. One reason offered for this is that the higher-order quantifiers are best seen as ranging over such genuinely higher-order entities, and that the hierarchy of types should be regarded as a hierarchy of the entities in the range of quantifiers of increasingly higher order: first-order quantifiers range over type 0 entities, second-order quantifiers range over type 1 entities, and so on.

Remarkably, already in 1944 Gödel had envisaged the possibility of taking the theory of types to be dealing with concepts or properties rather than classes. He writes:

It would seem that all the axioms of ***Principia***, in the first edition, (except for the axiom of infinity) are in this sense analytic for certain interpretations of the primitive terms, namely if the term 'predicative function' is replaced either by 'class' (in the extensional sense) or (leaving out the axiom of choice) by 'concept', since nothing can better express the meaning of the term 'class' than the axiom of classes (cf. page 140) and the axiom of choice, and since, on the other hand, the meaning of the term 'concept' seems to imply that every propositional function defines a concept. (Gödel 1944: 139)

---

[17] Note that these questions can be made sense of even from an anti-exceptionalist perspective. For conceptual naturalness can itself be considered a theoretical virtue.

Let us focus here on the property-theoretic interpretation. On this interpretation, we have objects at type 0, properties of objects at type 1, properties of properties of objects at type 2 and so on. Now on the type-theoretic conception of the universe, the types are incommensurable. As Quine pointed out, this is an *ontological* division. But, at least on an immediate reading, when the types are collapsed this ontological division is removed: properties (of whatever order) are now objects, entities in the first-order domain. Thus, on this reading, NF becomes a theory of *properties* and ∈ becomes a predication relation, by which a property can be predicated of other objects: $x \in y$ is to be read as *x has property y*.

Theories that treat properties as objects have been devised, for instance, by Peter Aczel (1980) and George Bealer (1982). However, there is a longstanding tradition which holds that it is simply incoherent to treat properties as objects, whatever the nuisances of typing. The tradition goes back at least to Frege's (1892) distinction between concept and object. According to him, an object is whatever a singular term refers to, and a concept is whatever a predicate refers to. But, he argued, singular terms and predicates cannot co-refer, and hence no concept (or property, in our case) is an object.[18] It is therefore 'a mere illusion to suppose that a concept can be made an object without altering it' (Frege 1953: x). Others have argued that whilst perhaps not incoherent, the thesis that properties are objects is not supported by the linguistic data (see, e.g., Chierchia and Turner 1988: §5.2).

However, these views are compatible with the idea that it is possible to systematically *associate* properties with certain objects. Indeed, this is precisely what Frege attempted to do with his infamous Basic Law V. This suggests an alternative reading of what happens when the types are collapsed. On this reading, the ontological division between objects and properties is preserved, in line with the Fregean strictures: the universe is still divided into objects, properties, properties of properties and so on. But when the types are collapsed, a property – a higher-order entity according to the type restrictions – is associated with an object: an object is taken as a *proxy* for a property. We say that the property has been *objectified*, but note that an objectified property is not, *sensu stricto*, a property. There will typically be more objects than those with which properties are associated – those will be, in the current context, the *Urelemente*, objects that are not objectified properties.

This process of objectification is the ontological counterpart to the linguistic process of *nominalization*, whereby we devise expressions which purport to refer in nominal positions to properties, which on the type-theoretic conception are referred

---

[18] This is also known as Frege's Concept *Horse* Paradox. See Trueman 2015 for a recent attempt at a reconstruction of Frege's argument.

to by means of predicates. An example will help (see Chierchia and Turner 1988: 264–265). According to the type-theoretic conception, *runs* in *Mary runs* picks out a property. However, *runs* is morphologically related to *running* as in *running is fun*. If properties and predicates cannot co-refer, then *running* cannot refer to a property. Being in nominal position, it refers to an object. This object is correlated with the property denoted by *runs*. Once the process of objectification is completed, we can also quantify in a first-order manner over entities that are suitably correlated with properties: we have *first-orderized* property talk.

According to the *stratified conception of set*, then, sets are objectified properties. They are what is obtained when higher-order entities in the type-theoretic hierarchy are systematically associated with objects. This can be seen as what is happening, from a conceptual point of view, when the types are collapsed. However, care needs to be taken when properties are objectified, for familiar Russellian reasons. For suppose that we take it that for every property there exists its objectification. This means that every second-order monadic property determines a set. That is, the principle (Collapse) from Section 1.7 holds:

$$\forall X \exists y \mathrm{Det}(X, y) \qquad\qquad \text{(Collapse)}$$

And recall that second-order logic includes the Axiom Schema of Comprehension (Comp2), stating that every condition determines a property. But as we saw in Chapter 1, (Collapse) and (Comp2) jointly imply the existence of the Russell set.

However, according to the stratified conception, the process of objectifiction should be restricted to those properties that are expressible by stratified conditions. That is to say, whilst to every condition there corresponds a property and so (Comp2) holds, the only properties that have objectifications are those expressible by a stratified condition. This means that (Collapse) needs to be restricted accordingly. Nino Cocchiarella (1985: 24–28) has shown that a natural development of this strategy leads to a system which contains and is equiconsistent with NF. Similarly, Cocchiarella shows that formulating (Collapse) so as to encompass *Urelemente* leads to a system which contains and is equiconsistent with NFU.[19] There remains an issue as to whether we should insist on Extensionality (or weak forms thereof) in the context of the stratified conception. For it is not obvious that we should associate the same objectified property to different albeit coextensive

---

[19] Another option is to restrict (Comp2) to stratified conditions and keep (Collapse) in full generality. Cocchiarella (1985: 13–24) shows that in this case too, the resulting system contains and is equiconsistent with NF (and similarly for the case with *Urelemente*). A similar result was later obtained by Hossack (2014) in the context of plural logic. He shows that a plural version of (Comp2) restricted to stratified conditions together with a version of full (Collapse) allowing for *Urelemente* naturally leads to a system which, when supplemented with Weak Extensionality, contains and is equiconsistent with NFU.

properties. If we do not, we end up with a system which has the same consistency strength as NF without Extensionality (which, in turn, has the same consistency strength as NFU, as noted earlier).

But how can the restriction to properties expressible by stratified conditions be motivated? Recall that according to the stratified conception, sets are objectified properties – they are proxies for properties. We are *implementing* properties as objects. However, whether an object satisfies an unstratified condition depends not only on what the object is like before the implementation, but also on the details of the implementation, on what the implementation looks like. But the following seems a plausible *desideratum* on the implementation: our implementation of properties with particular objects ought to be insensitive to the particular choice of association between objects and properties. Should we identify a property P with a different object in the first-order domain, P would still hold of the same things. The reason for this *desideratum* is that a particular association of properties with objects is *arbitrary*: there is no reason for thinking of an object as a proxy for a certain property rather than another one. The *desideratum* is reminiscent of the requirement of implementation independency familiar from computer science. According to this requirement, the implementation of abstract data types in a certain programming language should be as independent as possible of the details of the implementation (e.g. the peculiarities of the chosen programming language).

The stratified conception contends that given this *desideratum* on objectification (or, in computer science terms, on implementation of properties as objects), the conditions that pick out properties that have objectifications are precisely the ones that are stratified. For properties picked out by unstratified conditions may be had by an object only in virtue of the details of the objectification process; properties picked out by stratified conditions, on the other hand, are such that they apply to an object independently of which object they are associated with by the objectification process. Let us consider a couple of examples. Consider first the condition $x = x$. The property picked out by this condition is had by an object independently of the objectification/implementation process. Accordingly, it is a property that we should expect to be objectified. But now take the condition $x \in x$. If an object a satisfies it, it might well be due to the details of the implementation process. For instance, suppose that a is the proxy for the property P. If a were instead taken to be proxy for a different property, we need not have $a \in a$. This is compatible with the fact that $V \in V$, since regardless of which object $x = x$ is associated with, that object will have the property of being self-identical and hence belong to itself. The fact that $V \in V$ is just a side effect of the fact that V is self-identical, like any other object.

These are just particular cases. To make sure that they generalize appropriately, we need to make the notion of implementation independence mathematically precise. To this end, we use the notion of permutation invariance. Consider a set-theoretic interpretation $\langle \mathcal{M}, \in \rangle$ and let $\mathcal{M}^\sigma = \langle \mathcal{M}, \in_\sigma \rangle$. If $x \in_\sigma y$ just in case $x \in \sigma(y)$, then σ is a *permutation* of $\mathcal{M}$. We now say that a formula φ of $\mathcal{L}_\in$ is $\in$-*permutation invariant* just in case if a model $\mathcal{M}$ satisfies φ, then so does $\mathcal{M}^\sigma$. That is to say, a formula is $\in$-permutation invariant if the class of its models is closed under setlike permutations of the membership relation. Remarkably, a theorem of Pétry, Henson and Forster (Forster 1995: Theorem 3.0.4) says that the formulae that are $\in$-permutation invariant are exactly the stratified ones. This means that the stratified formulae are exactly those that denote properties which are implementation independent: which objects they hold of does not depend on the particular choice of association of properties with objects carried out in the objectification process. The stratification requirement, far from being ad hoc, turns out to be naturally motivated by the idea that sets are objectified properties.

6.7 NF As a Theory of Logical Collections

In the previous section, we introduced the stratified conception and showed that when sets are conceived of as objectified properties, the restriction to stratified conditions in the Naïve Comprehension Schema can be naturally motivated. This means that the stratified conception has a claim to being the conception lying behind NF and NFU. On the face of it, this means that the stratified conception needs to be considered a rival to the iterative conception of set. However, the possibility for the two conceptions to exist alongside each other might be brighter than it might seem at first sight.

Recall that according to the stratified conception, sets are objectified properties expressible by stratified conditions. Thus, according to this conception, sets are collections associated with properties or concepts. They are extensions, with the possible sole exception that although extensions are usually taken to be extensional entities, objectified properties need not be. This makes the stratified conception of set a *logical* conception of collection (see Section 1.8). Membership in a stratified set is determined by the property that the set is associated with: an object a belongs to an objectified property b just in case a has the property associated with b. As a result, each property divides the universe of objects into two sets: those which have the property and those which do not.

Given his rejection of properties and other intensional entities as 'creatures of darkness' (1956: 180), it is no surprise that Quine never considered the possibility of regarding NF as a theory of objectified properties. As Cocchiarella notes,

Quine rejected the whole idea of predication as something different from membership and replaced Russell's higher-order theory of predication by a first-order theory of membership, which he has continued to describe as a version of set theory, but which he no longer considers to be part of logic. (Cocchiarella 1992: 385)

Indeed, Quine seems to have rejected the distinction between logical and combinatorial collections altogether:

A class in the useful sense of the word is simply a property in the everyday sense of the word, minus any discrimination between coextensive ones. (Quine 1987b: 23)

But we need not follow Quine here. If we accept that there is a sensible distinction between a logical and a combinatorial conception of a collection, this opens up the way for regarding the stratified conception as existing alongside the iterative conception.

According to this proposal, the sets – the entities that we use in our foundations for mathematics – are provided by the iterative conception. This conception is often taken to be, and certainly can be spelled out as, a combinatorial conception of collection. By contrast, objectified properties – the entities that we use in the process of nominalization – are provided by the stratified conception. This conception is a logical conception of collection.

This distinction between objectified properties (also referred to as 'classes' or 'logical classes') on the one hand and iterative sets on the other is familiar and has often been made in the literature. For instance, Martin writes:

[S]ets are generated by an iterative construction process. Classes are given all at once, by the properties that determine which objects are members of them [...] 'set' is a mathematical concept and 'class' is a logical concept. (Martin, cited in Maddy 1983: 119)

Similarly, Parsons (1974: 8) says that it is 'at least not *obvious* that *extension* and *set* are just one concept'. Here, extensions are logical collections whilst sets are combinatorial ones. Parsons goes on to argue against the reducibility of sets to extensions, without taking a final stand on the reducibility of extensions to sets because of issues having to do with the possibility of unrestricted quantification over all sets.

Be that as it may, there is certainly a coherent position which takes the concept of objectified property and the concept of a set to be distinct concepts, differing primarily in the fact that the former is a concept of a logical collection, the latter is a concept of a combinatorial collection. If that is true, and if the stratified conception is best regarded as a conception of objectified properties, i.e. extensions, it seems possible for the NF and NFU collections to exist alongside iterative sets (with perhaps the issue of whether we must insist on objectified properties being extensional).

Recall that Gödel had said that, by distinguishing concepts from sets

we acquire not only a fairly rich and understandable set theory but also a clear guidance for our search for axioms that deal with concepts generally. We can examine whether familiar axioms for sets have counterparts for concepts and also investigate whether earlier attempts (e.g. in terms of the lambda-calculus and of stratification, etc.), which deal with sets and concepts indiscriminately, may suggest axioms that are true of concepts generally. (Wang 1996: 277)

In later remarks, however, Gödel expressed doubts about the suitability of the technique of stratification to serve as a foundation for a theory of concepts. In a comment to Wang, he observes:

Once we distinguish concepts from sets, the older search for a satisfactory set theory gives way to a similar search for a satisfactory concept theory. For this purpose, however, Quine's idea of stratification [1937] is arbitrary, and Church's idea [1932–1933] about limited ranges of significance is inconsistent in its original formulation. (Gödel as reported in Wang 1996: 278)

We have seen, however, that there is a natural motivation for the stratification requirement, based on general considerations about when some entities are taken to be proxies for some other entities.

Regarding NF and NFU as theories of objectified properties or classes, which are logical collections also provides an elegant explanation of how the paradoxes arise. In Section 3.1 we said that one way of explaining the appeal of the Naïve Comprehension Schema is as arising from a conflation of two different concepts, the concept of set and the concept of objectified property. We now see that this explanation fits very well with taking NF and NFU to be theories of objectified properties. For these theories block standard reasoning leading to the set-theoretic paradoxes by rejecting indefinite extensibility. This enables them to retain universality. By contrast, the iterative conception deals with the paradoxes by rejecting universality whilst retaining indefinite extensibility. Conflating sets and objectified properties, and hence operating with a concept of collection satisfying both indefinite extensibility and universality, leads to disaster, at least classically.

6.8 The No Intuitive Model Objection

So, the NF universe is what is obtained when the type-theoretic hierarchy is collapsed. But what reason is there to think that the result is a consistent theory? Why think that we can indeed associate each property described by a stratified condition with an object? As Frege himself noticed after the discovery of the inconsistency in the system of *Grundgesetze*, the simple existence of nominalization in natural language gives us no guarantee that each property can be associated with an object. 'An actual proof', he wrote, 'can scarcely be furnished' (1979: 182). Instead, he thought, we need to assume the existence of the said correlation as an unprovable law. But as he noticed 'if it was possible for there to be doubts previously, these doubts have been reinforced by the shock the law has sustained from Russell's paradox' (1979: 183).

These doubts can be partly mitigated by our diagnosis of the paradoxes and the way NF deals with them. For, as pointed out earlier, NF's solution is to reject indefinite extensibility at the expense of universality. And this does give us some reason that familiar paradox-inducing reasoning cannot be reproduced within the theory, since according to our diagnosis that reasoning is made possible by the presence of both indefinite extensibility and universality.

However, one might ask for more: one might want the conception of set behind NF to provide an intuitive model for the theory, in the same way as the iterative conception provides an intuitive model in the cumulative hierarchy for theories like ZFC (see also Section 4.5). And this the stratified conception does not seem to do. Thus, the first objection to the stratified conception is that it does not provide us with a picture of what the NF universe looks like. This, the objection might continue, is just the result of the fact that the stratification requirement is syntactic, even if it has a semantic counterpart (as shown by the fact that the class of stratified formulae coincides with he class of formulae that are ∈-permutation invariant).[20]

The best we can do is to start from type theory and see, within type theory, that we can carry out Jensen's model-theoretic argument establishing the consistency of NFU – we can show in type theory that if type theory is consistent, then so is NFU. Thus, one can start from type theory and show, from within type theory itself, that collapsing the types – now understood philosophically as objectifying properties – cannot introduce any inconsistency, at least if we do not insist on Extensionality (as opposed to Weak Extensionality). Note, moreover, that if we allow type theory to admit of infinite types, the consistency proof for NF given by Holmes can also be carried out within type theory. So, if correct, this proof establishes that we do not need to give up Extensionality in favour of Weak Extensionality.

Thus, the following picture emerges. We start with the theory of types, which does have an intuitive model in the type-theoretic hierarchy. Armed with the theory of types, we can then use Jensen's model-theoretic argument to establish that the

---

[20] This objection seems to tie in with the critical remarks we saw earlier that NF is not 'the axiomatization of an intuitive concept' and that there is 'no mental imagery which leads' to it.

reasons for regarding type theory to be consistent are also reasons for regarding the result of the collapse of type-theoretic hierarchy to be consistent (at least if we admit *Urelemente*). Is this sufficient to deal with the no intuitive model objection?

The answer seems to depend on whether we take NF to be a foundation for the whole of mathematics, in the way set theory aspires to be. If we do, then it does not seem legitimate to appeal to type theory to persuade ourselves of the consistency of NF. For such an appeal requires us to justify the axioms of type theory, thus running afoul of the requirement that a foundation for mathematics should be justificatory autonomous – that it should be possible to motivate and justify its claims without appealing to another theory. We shall return to the requirement of justificatory autonomy in Section 7.9, when discussing the requirement of autonomy for a foundation of mathematics more generally.

However, note that if we take NF(U) to be a theory of objectified properties, the autonomy requirement with regard to the consistency of NF(U) may no longer apply. For a theory of objectified property need not serve as a foundation for mathematics (which is instead provided by set theory) and hence we can rely on some other theory – such as type theory – to become convinced that NF(U) is indeed consistent. Thus, whilst the no intuitive model objection might be worrisome for the stratified conception of *set*, it loses its bite when targeting the stratified conception of objectified property.

6.9 The Conflict with Mathematical Practice Objection

One of the standard objections against NF is that it invalidates AC, which is central to mathematical practice. As anticipated earlier, defenders of NF have typically offered two responses, possibly complementary.

The first is to take AC to hold for cantorian sets, which is compatible with Specker's result that V cannot be well-ordered. This, it is argued, is sufficient to vindicate AC's role within mathematical practice. On this view, the failure of AC concerns those sets that standard set theory refuses to countenance. The standard view that AC holds is analogous to a view that takes no mammals to lay eggs because it refuses to countenance the existence of the platypus.

The second response is to point out that AC may be consistently added to NFU: as long as we are prepared to countenance *Urelemente*, we can take Choice to hold unrestrictedly, even for non-cantorian sets.

Both responses are satisfactory as far as they go, but they do raise issues. The first response appears to make the sets for which AC does not hold superfluous: if, in order to account for ordinary mathematics we only need cantorian sets, one might worry, we need not countenance the non-cantorian sets after all. Moreover, the first response does not tell us anything about why one might have expected AC to hold at all. The second response is limited in scope: in mathematical practice we want to be able to restrict attention to pure sets, whereas the failure of AC in NF with standard Extensionality seems to prevent us from doing just that (unless we take objectified properties not to satisfy Extensionality).

Taking NF to be a theory of objectified properties deals with these issues. The fact that AC may fail for non-cantorian objectified properties, thus making these collections unsuited from a foundational point of view, is no longer problematic, since this is not what these collections are invoked to do. Rather, they are invoked to, for instance, provide the semantic value of nominalizations of predicates. This role does not require them to satisfy AC. Moreover, it is because we thought of NF as a theory of combinatorial collections that we were led to thinking that AC was to be expected to hold. But we should not expect NF, as a theory of logical collections, to validate AC. As we have seen, it is the combinatorial conception of collection that seems to lend some support to AC. Finally, all of this holds regardless of whether we restrict attention to pure sets or not.

6.10 Conclusion

The standard view that NF does not embody a conception of set is correct in some respects but not in others. It is correct in that there does not seem to be a conception of set behind NF which gives rise to an intuitive picture of the set-theoretic universe. But it is not correct in that the stratified conception – best thought of as a conception of objectified property – can be taken as backing NF. Whilst not delivering an intuitive picture of the NF universe, this conception motivates the stratification requirement, which has long appeared as a mere technical trick, in a principled manner.

Moreover, if NF is taken to embody a conception of objectified properties rather than sets, the standard objections to NF – its lack of an intuitive model and the fact that it contradicts the Axiom of Choice – lose their bite. This opens up the possibility for regarding the stratified conception as a conception of collection that can exist *alongside* the iterative conception of set.

What is more, once this is done, we have a simple and powerful explanation of the paradoxes: they arise from the conflation of two different conceptions of collections. The first conception – the stratified conception – is a conception of objectified property (or class) and hence a logical conception of collection; it satisfies universality but rejects indefinite extensibility. The second conception – the iterative conception – is a conception of set, which can be naturally articulated as a combinatorial conception; it satisfies indefinite extensibility but rejects universality. The conflation of sets and objectified properties is disastrous in that it leads, provably, to inconsistency.

Thus, NF is best regarded as a theory of objectified properties. As such, however, it should be evaluated against other theories of the same subject matter and the conceptions they embody (see, e.g., Field 2008; Schindler 2019).[21]

---

[21] For reasons of space, we have not considered Church's Universal Set Theory and related theories, which constitute a class of theories with a universal set different from NF(U). Unlike the current situation with regard to NF(U), these theories seem to be backed by a conception which provides us with an intuitive model of them (see Forster 2008). Very briefly, the conception lying behind these theories, which Forster calls a kind of iterative conception, consists in forming, at each stage of the iteration process, the sets standardly occurring at a level of the standard cumulative hierarchy as well as their complement. The conception has many attractive features but appears less simple than the iterative conception and disjunctive in character without leading to a theory with substantial mathematical advantages over iterative set theory from a foundational point of view. The iterative conception, therefore, seems to fare better than this conception.

7

The Graph Conception

As to classes in the sense of pluralities or totalities
it would seem that they are likewise not created
but merely described by their definitions and that therefore
the vicious circle principle in the first form does not apply.
I even think there exist interpretations of the term 'class'
(namely as a certain kind of structures),
where it does not apply in the second form either.

Kurt Gödel (1944: 131)

The iterative conception's explanation of the paradoxes takes the concept of set to be indefinitely extensible, whilst giving up its universality. On the iterative conception, in particular, there is no set of all sets.

But as well as giving up universality and hence banning sets like the universal set, the iterative conception also holds that all sets occur at some level or another of the cumulative hierarchy, and this rules out non-well-founded sets. It is precisely to guarantee that the universe of sets has this well-founded structure that iterative set theories include axioms such as Foundation or axiom (ρ).[1]

Now, clearly, the conceptions we encountered in Chapters 4 and 6 respectively, that is, the naïve conception and the stratified conception, allow for the existence of non-well-founded sets. For these conceptions do *retain* universality and hence sanction the existence of the non-well-founded set par excellence, namely the set of all sets. But if the diagnosis of the paradoxes offered in Chapter 2 is correct, it

---

[1] However, we saw in Chapter 2 that in the context of the axioms of Z the Axiom of Foundation, contrary to axiom (ρ), does not deliver the hierarchy even if we switch to the second-order level. Of course, in the first-order case, the set theories which result from the adoption of these axioms will still admit of models which contain non-well-founded sets. Nonetheless, these theories are intended to capture a picture of the set-theoretic universe as a well-founded structure.

should be possible to block the paradoxes by banning universality whilst admitting non-well-founded sets.

To the extent that ZFC itself is consistent, this has been rigorously shown by Aczel (1988). In his book, he considers and develops four set theories which reject the existence of the set of all sets or the set of all ordinals whilst admitting non-well-founded sets.[2] Moreover, these set theories are all consistent if and only if ZFC is – that is, given a model of ZFC, we can construct a model of each of these theories and vice versa (see Aczel 1988: chs. 5–7). Since these particular non-well-founded set theories will be our focus in the remainder of this chapter, I shall often refer to these theories simply as 'non-well-founded set theories'; the context will make it clear when what I mean is *non-well-founded set theories of the kind considered by Aczel*.

Non-well-founded set theories, and in particular the one based on the axiom AFA (to be described below), have received a wealth of attention by mathematicians (especially category theorists) and computer scientists, but have been largely neglected by philosophers.[3] At the root of this neglect may lie the impression that non-well-founded set theories do not embody a conception of set, but are rather of mere technical interest. In this chapter we dispel this impression, at least for the system based on AFA (or, perhaps, some subsystem thereof, as we shall see).

The plan of the chapter is as follows. First, I review the key ideas of Aczel's approach, and present the four systems he considers. Then, I describe what I shall call the *graph conception of set*, a conception of set which may be taken as lying behind a non-well-founded set theory. In the light of this conception, I reassess Adam Rieger's (2000) claim that if we admit non-well-founded sets at all, we should admit more than those in the AFA universe. I argue that AFA is justified on the graph conception, which provides, *contra* Rieger, a *rationale* for restricting attention to the system based on this axiom. I then turn to the other axioms of this system, and attempt to make a case that most of them are justified on the graph conception too. My case will rest on informal considerations and on the possibility of showing that some of these axioms follow from a theory formalizing part of the content of the graph conception. I conclude by discussing four objections to the graph conception and comparing the iterative conception and the graph conception in the light of this discussion. Some morals are drawn about the significance of the graph conception for the debate about the justification of the Axiom of Foundation.

---

[2] These theories are pure, and hence in keeping with the restriction mentioned in Section 2.1. We discuss the issue of how to give impure versions of these theories in Section 7.8.

[3] For some use and discussion of non-well-founded sets in mathematics and computer science, see, e.g., Turi and Rutten 1998, Rutten 2000, Johnstone et al. 2001 and van den Berg and De Marchi 2007. Barwise and Etchemendy 1987 is a rare attempt to put non-well-founded set theory to philosophical use.

7.1 Depicting Sets with Graphs

All non-well-founded set theories are obtained by adding to ZFC⁻ (ZFC minus the Foundation Axiom) some version of an *anti*-foundation axiom.[4] What all of these versions of the anti-foundation axiom have in common is that they make use of the idea of a set being depicted by a *graph*.

Typically, a graph consists of some points possibly connected by some lines; an example is provided in Figure 7.1. The points are generally referred to as the *nodes* of the graph; the lines, or whatever connects the points, as its *edges*. For this reason, in standard set theory a graph is usually defined as consisting of two sets: the set of the nodes of the graph, and the set of its edges – where each edge is a pair of nodes. We may use graphs to depict sets by taking nodes to represent sets, and edges to represent (converse) membership. But for this to work, we need to place certain restrictions on the kind of graphs we use.

To begin with, since edges are to represent converse membership, we want them to have a single, designated direction. Thus, we restrict attention to *directed* graphs – graphs in which every edge has such a direction. Graphically, this is indicated by using arrows as edges; set-theoretically, it is captured by taking edges to be *ordered* pairs of nodes. An arrow from a node n to a node n' will then mean that the set represented by n' is a member of the set represented by n. We write $n \longrightarrow n'$ to indicate the presence of such an arrow, and say that n' is a *child* of n.

Given a graph, we want to be able to locate the set it depicts. To this end, we demand that a graph, besides being directed, be also *pointed* – that it have a unique, distinguished node, called the *point*. The point represents the set depicted by the

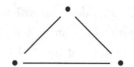

Figure 7.1 An undirected graph.

---

graph, and is usually shown at the top of the picture. (For this reason, it is sometimes called the *top node* of the graph.) What about the other nodes of the graph? They represent sets in the transitive closure of the set represented by the point. So let a *path* be a finite or infinite sequence

$$n_0 \longrightarrow n_1 \longrightarrow n_2 \longrightarrow \ldots$$

of nodes $n_0, n_1, n_2, \ldots$ each of which is a child of its predecessor. Then, we require that the graph be also *accessible*, i.e. that it be possible to reach each node of the graph by some finite path starting from the point. If this path is always unique, the graph is said to be a *tree*, and the point is said to be its *root*.

Thus, we use (directed) accessible pointed graphs (*apgs* for short) as pictures of sets. (And, as the context will make clear, we shall often take the word 'graph' to mean *apg*.) But when does an apg depict a certain set? To answer this question, we need to introduce some further terminology. A *decoration* of a graph is an assignment of elements of the set-theoretic universe to each node n of the graph such that the elements of the set assigned to n are the sets assigned to the children of n. It follows that a childless node must be assigned the empty set. More generally, one can use the Axiom of Replacement to show that in the case of well-founded graphs – that is, graphs which have no infinite path – it is determined which elements of the universe are assigned to nodes: every well-founded graph has a unique decoration (Aczel 1988: 4–5). A *picture* of a set a is then an apg which has a decoration which assigns a to the point; if in addition the decoration assigns distinct sets to distinct nodes, then the apg is said to be an *exact* picture of a.

Since every set has, up to isomorphism, one and only one exact picture,[5] we can know which sets there are if we know which apgs are exact pictures. Each non-well-founded set theory gives a different answer to this question, and the role of the anti-foundation axiom can be seen as precisely that of specifying which apgs are exact pictures. Our next task is to illustrate these theories and the corresponding anti-foundation axioms.

7.2 Four Non-Well-Founded Set Theories

Say that a graph is *extensional* if and only if no two distinct nodes have the same children. The first theory, *Boffa* set theory, is obtained by adding to ZFC$^-$ Boffa's

---

[5] Any doubt about this claim can be dispelled by considering the notion of a *canonical picture* of a set a. This is the apg whose nodes are the sets that occur in sequences a_0, a_1, a_2 such that $\ldots a_2 \in a_1 \in a_0 = a$, whose edges are pairs $\langle a, b \rangle$ such that $a \in b$, and whose point is simply a. Clearly, each set has exactly one canonical picture. But another definition of an exact picture, equivalent to the one we are using (see Aczel 1988: 28), tells us that an apg is an exact picture iff it is isomorphic to a canonical picture. The claim follows at once.

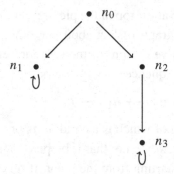

Figure 7.2 An exact picture of a Boffa set.

Anti-Foundation Axiom, or **BAFA** for short (see Boffa 1969), which has the consequence that any extensional apg is an exact picture.[6]

It is usually claimed, however, that if our guide to set existence is given by
graphs, there should be further constraints on a graph being an exact picture besides
it being extensional. A typical example is offered by the graph shown in Figure 7.2
(see Rieger 2000: 245). In Boffa set theory, this is an apg depicting a set a such
that $a = \{b,c\}$, $b = \{b\}$, $c = \{d\}$ and $d = \{d\}$. However, a number of writers (e.g.
Rieger 2000) have felt that if our guide to sethood is provided by apgs, the sets b
and d should be identical.[7] (Notice that the Axiom of Extensionality will be of no
use here, since it simply leads to the conclusion that $b = d$ if and only if $b = d$.)

To deal with cases of this sort, it is customary to introduce the notion of
isomorphism-extensionality. To do this, however, we first need to introduce the
notion of an induced subgraph. Intuitively, an *induced subgraph* (or simply a
subgraph) $\mathcal{H}$ of a graph $\mathcal{G}$ will consist of some nodes taken from $\mathcal{G}$ and all edges
of $\mathcal{G}$ between these nodes. In set theory, this is captured by saying that $\mathcal{H}$ is a
subgraph of $\mathcal{G}$ if and only if the set of nodes of $\mathcal{H}$ is a subset of the set of nodes of
$\mathcal{G}$ and for any two nodes n_1, n_2 of $\mathcal{H}$, $\langle n_1, n_2 \rangle$ belongs to the set of edges of $\mathcal{H}$ if
and only if $\langle n_1, n_2 \rangle$ belongs to the set of edges of $\mathcal{G}$. A graph $\mathcal{G}$ is then said to be
isomorphism-extensional if and only if there are never distinct nodes n, n' of $\mathcal{G}$ such
that the subgraph below n – i.e. the subgraph having n as point and consisting of all
nodes lying on paths starting from n – is isomorphic to the subgraph below n'. With

---

[6] In his book, Aczel briefly considers a related axiom also due to Boffa, *viz.* Boffa's Weak Axiom BA_1, which is
equivalent to the assertion that an apg is an exact picture iff it is extensional. Clearly, then, BAFA implies BA_1,
but the converse is not true. See Aczel 1988: 57–65 for details.

[7] Thus, concerning a very similar example, Rieger (2000: 244–245) writes that 'there is an intuition that [the two
sets] *ought* to be equal. There cannot, it seems, be any good reason for distinguishing them – and certainly not
any reason which did not have an intensional feel to it'.

this notion on board, we can formulate one version of Finsler's Anti-Foundation Axiom (FAFA), which asserts that an apg is an exact picture of a set if and only if it is extensional and isomorphism-extensional. *Finsler-Aczel* set theory[8] is then the result of adding FAFA to the axioms of ZFC⁻.

The third theory builds on the work of Scott (1960), and is based on the idea that, given an apg, we can unfold it to obtain a tree which depicts the same set. In particular, the *unfolding* of an apg is the tree whose nodes are the finite paths of the apg which start from its point n_0 and whose edges are pairs of paths of the form

$$(n_0 \longrightarrow \ldots \longrightarrow n, n_0 \longrightarrow \ldots \longrightarrow n \longrightarrow n').$$

The root of the tree is the path n_0 of length one. Given a decoration D of the apg, we can obtain a decoration of its unfolding by assigning to the node $n_0 \longrightarrow \ldots \longrightarrow n$ of the tree the set that is assigned by D to the node n of the apg. It follows that the unfolding of an apg will depict any set depicted by the apg. Now say that an apg $\mathcal{G}$ is *Scott-extensional* if and only if there are never distinct nodes n, n' of $\mathcal{G}$ such that the subtree below $n_0 \longrightarrow \ldots \longrightarrow n$ in the unfolding of $\mathcal{G}$ is isomorphic to the subtree below $n_0 \longrightarrow \ldots \longrightarrow n'$.[9] *Scott* set theory is then obtained by adding to ZFC⁻ Scott's Anti-Foundation Axiom (SAFA), which is equivalent to the claim that an apg is an exact picture if and only if it is Scott-extensional.[10]

To appreciate the effects of this axiom, consider the apg $\mathcal{G}$ with point n_0 in Figure 7.3 (see Aczel 1988: 54–55). A moment's reflection shows that this apg is extensional and isomorphism-extensional. On the other hand, if we label each node $n_0 \longrightarrow \ldots \longrightarrow n$ of the unfolding of $\mathcal{G}$ with the name of the node n of $\mathcal{G}$, we obtain the diagram in Figure 7.4. And this diagram makes it clear that the subtrees below n_1-labelled and n_2-labelled nodes are isomorphic. Thus, $\mathcal{G}$ is not Scott-extensional, and is therefore an exact picture in Finsler-Aczel set theory but not in Scott set theory. In particular, if FAFA is assumed, there is a decoration of the graph which assigns pairwise distinct sets a, b, c to the nodes n_0, n_1, n_2 such that $a = \{c\}$, $b = \{a, c\}$ and $c = \{a, b\}$. If SAFA is assumed, on the other hand, such a decoration is ruled out, and the only admissible decorations are ones that do not assign distinct sets to distinct nodes of the graph, namely the decoration in which

---

[8] The terminology is Rieger's (2000: 246), who motivates the choice by explaining how the theory is the result of applying original insights of Aczel's to the work of Finsler (1926).

[9] A *subtree* is a subgraph of a tree, and is itself a tree. Note that the latter follows, as desired, because we are restricting attention to apgs.

[10] Thus, whilst isomorphism-extensionality has to do with isomorphisms between sub*graphs*, Scott-extensionality has to do with isomorphisms between sub*trees*. Hence, Scott set theory can be seen as what one ends up with if one focuses on isomorphisms when trying to articulate the identity conditions of non-well-founded sets but takes trees, rather than apgs, to provide the primary guidance as to what sets there are.

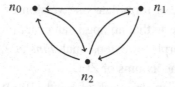

Figure 7.3 An exact picture of a Finsler-Aczel set.

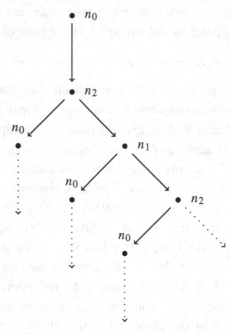

Figure 7.4 Unfolding of the apg with point n_0 in Figure 7.3.

the nodes n_1 and n_2 get assigned a set a and the node n_0 gets assigned a distinct set b such that $b = \{a\}$ and $a = \{a, b\}$, and the decoration which assigns the set $\mho = \{\mho\}$ to every node.[11]

The fourth set theory is ZFA, and is obtained by adding to the ZFC$^-$ axioms the Anti-Foundation Axiom (AFA), discovered independently by Forti and Honsell (1983) and Aczel. This axiom states that every graph has a unique decoration. As a result, every apg is a picture of exactly one set. The axiom is obviously equivalent to the conjunction of the two following statements:

Every graph has at least one decoration. (AFA$_1$)

Every graph has at most one decoration. (AFA$_2$)

---

[11] The standard symbol in the literature is 'Ω' but I am using '$\mho$' to avoid confusion with the symbol for the property of being an ordinal.

In other words, the axiom has an existence part and a uniqueness part. The uniqueness part has the consequence that even more decorations are ruled out than when SAFA is assumed. So, for instance, the only decoration of the graph in Figure 7.3 is the one where each node is assigned the circular set $\mho$. On the other hand, and as the same example also shows, the fact that the axiom has an existence part also means that it is still compatible with the existence of non-well-founded sets. In particular, in the presence of the ZFC⁻ axioms – which guarantee the existence of well- and non-well-founded graphs – the axiom implies the existence of non-well-founded sets alongside the well-founded ones, just as the other anti-foundation axioms do.

Like the other anti-foundation axioms, moreover, AFA too can be understood as specifying which apgs are exact pictures. Roughly speaking, the axiom can be seen as saying that a graph is an exact picture just in case it has no distinct nodes such that exactly the same movements are possible along the edges departing from these nodes. To make this rigorous, we need to introduce the notion of a bisimulation. We say that a relation R is a *bisimulation* on a graph $\mathcal{G}$ if and only if, for any two nodes n and n' of $\mathcal{G}$, whenever R holds between n and n', then for each child of n there is a related child of n' and vice versa. A relation R is then said to be the *largest bisimulation* on a graph $\mathcal{G}$ (written $\equiv_{\mathcal{G}}$) if and only if R is a bisimulation on $\mathcal{G}$ and whenever there is a bisimulation on $\mathcal{G}$ between two nodes n and n', then R holds between them. Now say that a graph $\mathcal{G}$ is *strongly-extensional* just in case for all n, n' in $\mathcal{G}$, if $n \equiv_{\mathcal{G}} n'$, then $n = n'$. The key result proved by Aczel (1988: 28) is then that AFA is equivalent to the statement that an apg is an exact picture if and only if it is strongly-extensional.

We can use the result to further illustrate the differences between Scott set theory and ZFA. Consider the apg $\mathcal{H}$ in Figure 7.5 (see Aczel 1988: 53–54). A look at its unfolding $\mathcal{H}'$ in Figure 7.6 (with nodes labelled employing the same induced labelling used for the unfolding in Figure 7.4) reveals that it is Scott-extensional, since no subtree of $\mathcal{H}'$ below an n_0-labelled node is isomorphic to a subtree below an n_1-labelled node. However, $\mathcal{H}$ is clearly not strongly-extensional. Thus, whilst

Figure 7.5 An exact picture of a Scott set.

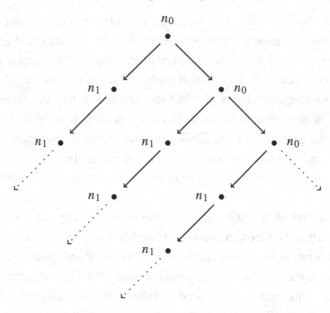

Figure 7.6 Unfolding of the apg in Figure 7.5.

an exact picture of a set $a \neq \mho$ such that $a = \{a, \mho\}$ in Scott set theory, $\mathcal{H}$ is simply a non-exact picture of $\mho$ in ZFA.

7.3 The Graph Conception of Set

Hence, we have four (presumably) consistent set theories which admit non-well-founded sets but which are more or less generous in the kind of non-well-founded sets whose existence they allow for. Moreover, all the standard mathematics that can be developed in ZFC can be developed in these theories.

However, one might think that these theories, albeit apparently consistent and powerful enough to embed standard mathematics, are simply fragmentary collections of axioms – there is no thought behind them, as they say. This, one might suppose, is in sharp contrast with the case of ZFC, which, together with its extensions and subsystems, does embody a conception of set, namely the iterative conception. Thus, one might conclude, non-well-founded set theories are of mere technical interest, and do not deserve any attention on the philosophers' part.

One might reason in this way, but one would be mistaken. For there *is* a conception of set which can plausibly be taken as lying behind a non-well-founded set theory. This conception of set is not at all ad hoc, and indeed arises quite naturally just by considering the idea that sets may be depicted by graphs. The conception, moreover, is very easy to state, and has the resources to motivate most of the ZFA axioms. I will introduce the conception in this section, and will investigate its relation to the four non-well-founded set theories in the next two.

We have seen that the fundamental idea of Aczel's approach is the idea of a set being depicted by a graph. By taking sets to be the kind of things that are depicted by graphs, non-well-founded sets arise quite naturally alongside the well-founded ones. We have graphs that have an infinite descending path, and, among those, graphs which contain loops; to all of these, there will correspond circular sets of one kind or another. Hence, when trying to articulate a conception of set motivating a non-well-founded set theory, a natural suggestion is to make the notion of a set being depicted by a graph central to the conception. This naturally leads to the *graph conception of set*, which states that sets are what is depicted by an *arbitrary* graph.[12]

Of course, to develop a workable theory of sets on the basis of this conception, there must be some constraints on what can count as a graph. As we have already seen, in developing a theory of non-well-founded sets we restrict attention to accessible pointed graphs. This restriction, however, is easily motivated: if a graph is to depict a set, then we want to be able to tell which set it depicts and which members this set contains. Furthermore, this is the *only* restriction that we want on what should count as a graph: *any* apg is a graph in good standing. It follows that according to the graph conception of set, to every apg $\mathcal{G}$, there corresponds a set which $\mathcal{G}$ depicts.

But, it may be asked, why should one even *begin* to think of sets as what is depicted by graphs? Here is one possible train of thought. Suppose you are told that a set is an object which may have members, the objects which bear the membership relation to it. And suppose, further, that you are told that the members of a set are themselves sets – recall that we are restricting attention to pure sets – and hence that they too may have members. You will then realize that *these* members will also be sets, and so that *they* too may have members, which will be sets and will therefore be capable of having members. More generally, you will realize that chains of membership can be rather long; in fact, you will note that you have been given no reason to think that they are always finite, and so you will make no assumption to this effect.

Now, since all you have been told is that a set is an object which may have members and that the members of a set are themselves sets, you might suppose that

12 Another possible route to a non-well-founded set theory is via a limitation of size conception. Thus, Jané and Uzquiano (2004: §5) show that there are models of FN* which satisfy AFA. What remains open, however, is whether these models can be considered intended models of FN*.

a set *simply* is an object of this kind – an object having a (hereditary) *membership structure*. That is to say, you might suppose that a set *just is* an object which may bear the converse of the membership relation to objects which, being sets, may bear this relation to sets, and so on. This supposition might be reinforced when you note that it is the structural richness of sets which enables us to employ them for the purposes you have heard we typically use sets for, such as embedding classical mathematics within set theory and performing various modelling jobs. For, you might then think, no interesting uses of sets are lost by taking a set simply to be an object having membership structure.

Now suppose that you come to see that any such structure, no matter how complicated, can always be fully represented by some graph. And suppose, moreover, that you realize that, if you restrict attention to apgs, what will be depicted by a graph will always be something having membership structure. Then, you will see that you can simply take sets to be precisely the things that are depicted by apgs, as the graph conception maintains. Note that since you have decided not to assume that all chains of membership are finite, you have accordingly not assumed that only graphs having no infinite path depict objects which have membership structure. To sum up, if one, quite naturally, thinks of sets simply as objects having membership structure, one can just take them to be what is depicted by graphs of the appropriate form. Note that the graph conception, so motivated, is naturally seen as a combinatorial conception of set, since sets are characterized by reference to their membership structure rather than to some condition, property or concept.

The idea that a set is simply an object having membership structure features in the work of Barwise and Moss (1991: 36–37; see also Baltag 1999: 482–493). They suggest that we think of a set as what one obtains from an apg by a process of abstraction: if we take a graph and forget the particular features of the nodes and edges, they claim, what we are left with is 'abstract set-theoretic structure'. Interestingly, an analogous suggestion was made much earlier by Mirimanoff, who writes:

What is common to isomorphic sets is the structure of their composition. If one abstracts from the particular properties which distinguish a set from its isomorphs, and if one retains only the details of its structure, one arrives at a new concept which one could call the structure type of this set, a concept closely related to Cantor's definition of power and ordinal number. In fact, it coincides with the notion of power or cardinal number when one abstracts from the structure of the elements of the set, which amounts to regarding these elements as urelements. On the other hand, the order-types of well-ordered sets, Cantor's ordinal numbers, are derived directly from the structure types of certain special sets which I have called *S* sets. (Mirimanoff 1917: 212, translation in Hallett 1984: 191)

Here, Mirimanoff is not suggesting that we think of sets as what one obtains from apgs by a process of abstraction. Rather, we start from sets, and through abstraction we arrive at structure types. All the same, Barwise and Moss's notion of an abstract set-theoretic structure is clearly anticipated by Mirimanoff's notion of a structure type.

Let us conclude this section by addressing a possible initial worry about the notion of a graph presupposed by the graph conception. Recall that in Section 7.1 we saw that a directed graph can be characterized as consisting of a set of nodes and a set of edges, the latter being order pairs of nodes. This might suggest that the graph conception rests on a prior notion of set, which would seem to jeopardize its ability to serve as a foundation for set theory. This objection fails on closer inspection, however. Although graphs can be characterized in set-theoretic terms, they need *not* be so characterized. The claim of the defender of the graph conception of set, presumably, is that we have an independent grasp of the notion of a directed graph which is not mediated by set theory. To be sure, it is not enough to say that we have a grasp of the notion of a directed graph which is not mediated by set theory; we must also have a grasp of other graph-theoretic notions, such as that of an apg, which is not so mediated. Again, however, the claim of the defender of the graph conception of set is going to be that although these notions can be captured in a set-theoretic framework, we can independently grasp them in non-set-theoretic terms. This is why we have first introduced notions such as that of a directed graph and that of a subgraph by way of examples and in terms of points connected by lines: introducing these notions in this way makes it more plausible that we can understand them without going through their set-theoretic definitions.

7.4 The Graph Conception and AFA

Among the non-well-founded set theories, nearly all attention has been directed to ZFA.[13] Aczel himself devotes most of his discussion to this theory, and refers to AFA as *the* anti-foundation axiom (1988: xviii). Despite the focus on ZFA, however, the literature is surprisingly short of arguments for AFA.[14]

---

[13] For instance, standard textbooks such as Devlin 1993 and Moschovakis 2006, when discussing non-well-founded set theory, focus on ZFA. Moreover, applications of non-well-founded set theory have mostly relied on AFA; in addition to the references provided in fn. 3 above, see Barwise 1986 and Barwise and Moss 1996, which make use of AFA to deal with issues in logic and linguistics.

[14] This is not to say that there is no explanation as to why people have directed most of their attention to ZFA. For one thing, the uniqueness part of AFA is very useful when non-well-founded set theory is used to deal with, e.g., streams and the semantic paradoxes. For another, the fact that the notion of bisimulation can be used to give an equivalent formulation of AFA reveals deep connections between ZFA and the theory of coalgebra, and these connections are central to many applications of non-well-founded set theory in category theory and computer

In its support, Aczel (1988: 4–6) invokes the fact that with the help of Replacement we can show that every well-founded graph has a unique decoration (see Section 7.1). The idea is that the uniqueness property in this case motivates the corresponding uniqueness in the AFA case. But, as Rieger (2000: 249) notices, this argument does not work. For it is precisely well-foundedness that enables us to define by recursion the unique function decorating the graph. Hence, uniqueness in the well-founded case gives us no reason to expect uniqueness in the non-well-founded case too.

Moschovakis (2006: 238–239) first attempts to motivate AFA in a similar but unsuccessful way,[15] and then goes on to say that the ' "uniqueness" part of the Antifoundation Principle [. . .] makes it possible to specify and analyze the structure of ill founded sets with diverse properties, and is the main advantage of the antifounded universe $\mathcal{A}$ [i.e. the non-well-founded universe generated by AFA] over other models which contain ill founded sets'. It is hard to understand what Moschovakis has in mind here. If he is saying that it is only possible to analyze the structure of non-well-founded sets when we restrict attention to the AFA universe, that seems false: in the other non-well-founded set theories, to certain graphs there will correspond more than one set; but it will still be possible to investigate the structure of these sets. If the point is that the structure of the AFA universe is easier to investigate, this might well be right, but it is hard to see why it constitutes an argument for AFA. After all, 'the cumulative hierarchy of the iterative universe has an enticingly elegant mathematical structure' (Aczel 1988: xviii) unsurpassed by any of the non-well-founded universes. Yet, we do not want to say that this alone motivates well-foundedness.[16]

The lack of good arguments for AFA in the literature has led Rieger to conclude that, just like the restriction to well-founded sets, the restrictions imposed by AFA are unmotivated. According to him, the only restrictions we should accept are those imposed in Finsler-Aczel set theory, which 'gives the richest possible universe of sets while respecting the extensional nature of sets' (2000: 247). To understand what Rieger means by this, recall that in the current framework we can know which

---

science. See Moss 2009, which argues that these connections suggest that the contrast between Foundation and AFA is just an instance of a more general division between 'bottom-up' and 'top-down' approaches, and explores the prospects for using this fact to provide a motivation for ZFA.

[15] 'Each grounded graph G admits a unique decoration d_G, and the pure, grounded sets are all the values $d_G(x)$ of these decorations. Can we also 'decorate' the nodes of ill founded graphs to get pure, ill founded sets which are related to ill founded graphs in the same way that pure, grounded sets are related to grounded graphs?' The answer, of course, is 'Yes', provided that AFA holds.

[16] Barwise and Moss (1996: 68–69) also offer a brief argument in favour of AFA. But even if successful, all the argument allows us to conclude is that isomorphic sets are equal, and so can at best motivate the axiom FAFA. See Rieger 2000: 249 for details.

sets there are if we know which apgs are exact pictures: the more generous the answer to the question 'When is an apg an exact picture?', the larger the universe of sets. BAFA, as we have seen, gives the most generous answer which does not violate the Axiom of Extensionality. AFA, at the opposite extreme, gives the most restrictive answer compatibly with the fact that every graph has at least one decoration. As a result, the universe of sets in ZFA is the smallest among those of the four non-well-founded set theories. But, Rieger claims, there is no reason to accept the restrictions imposed by AFA. The only restrictions we should recognize, he contends, are those required by FAFA, since in his view the extensional nature of sets demands that exact pictures, besides being extensional, be also isomorphism-extensional. Hence, he concludes, 'if we admit non-well-founded sets at all, we ought to admit more than there are in the AFA universe' (2000: 252).

The graph conception of set, however, provides a philosophically sound reason for mathematicians' focus on AFA. To see this, we need to make use of the fact that AFA is equivalent to the conjunction of AFA_1 and AFA_2: what we shall show is that these two statements are justified on the graph conception.

It is easy to see that AFA_1 is justified on the graph conception. For, as noted above, it follows from the graph conception that to every apg $\mathcal{G}$, there corresponds a set which $\mathcal{G}$ depicts, which sanctions the assertion that every graph has at least one decoration. What about AFA_2, the uniqueness part of the anti-foundation axiom AFA? There are two arguments to the effect that this too is justified on the graph conception.

The first argument is epistemological, and is based on the role that graphs are supposed to play in our theory of sets according to the graph conception. According to this conception, sets are what is depicted by an arbitrary graph. Hence, on this conception, graphs provide our only guide to what sets there are. The idea is that given a graph $\mathcal{G}$, we can move from $\mathcal{G}$ to the set it depicts. Earlier on, we made use of this idea in order to motivate the restriction to accessible pointed graphs when explicating the notion of an arbitrary graph which we are working with. The idea, we then pointed out, is that if a graph is to depict a set, we want to be able to tell which set it depicts and which members this set contains. The restriction, therefore, is motivated on the grounds that the role that graphs play in our theory of sets according to the graph conception is that of providing a guide to what sets there are, and for this to be possible, we need to be able to move from graphs to the sets they depict.

The idea that we should be able to move from graphs to sets, however, also provides an argument for AFA_2. For if any given apg depicts exactly one set, as AFA_2 in conjunction with AFA_1 implies, then, given a graph of the appropriate form, there is nothing else we need to know in order to know which set that graph depicts. If, on the other hand, to every graph there corresponds more than one set, we also need to know which particular decoration of that graph we are considering. This conflicts with the idea – crucial to the graph conception of set – that graphs are our only guide to what sets there are.

It is also worth noticing that set theorists, when introducing set theories of the kind we are considering, seem to take it for granted that it should be possible for us to move from a graph to the set it depicts without any additional information concerning the decoration of the graph. A clear example is provided by the following passage:

Given some structured object *a* in the world, we may (in theory, at least) represent its hereditary constituency relation by means of a graph and thereby obtain a 'set-theoretic' model of *a* by moving from the graph to *the* set it depicts – namely, the set that corresponds to the top node of the graph. (Devlin 1993: 150, my emphasis).

The passage is also interesting because it contains the suggestion that if we want to obtain set-theoretic models for structured objects in the world (such as items of information in some information-storage device – see Devlin 1993: 143), then it is easy to proceed by first representing the structure of these objects by means of a graph, and then moving from the graph to the set it depicts. This means that if one takes sets to be what is depicted by graphs, one can straightforwardly obtain set-theoretic models of structured objects which appear in the world and hence use sets to perform various modelling jobs.

The second argument that AFA_2 is justified on the graph conception is ontological, and is based on what the conception takes sets to be. The graph conception, to repeat once again, states that sets are what is depicted by an arbitrary graph – with the restriction that the only admissible graphs are the accessible pointed ones. The idea, in other words, is that sets are just the things that correspond to graphs of the appropriate form. But if that is true, then the identity of sets is determined by the graphs they correspond to. Another way of making the same point is to recall that one possible route to the graph conception is to think of sets as just objects having membership structure. Since the membership structure of any set can be fully represented by some apg, the thought was, we can simply take sets to be the things that are depicted by apgs. These will then represent the membership structure of the sets they depict. But if apgs represent the membership structure of the sets they depict, they should also decide questions of identity between them. For if sets are simply objects having membership structure, then this structure should determine their identity.

We see, therefore, that the following analogue of the Axiom of Extensionality holds on the graph conception: two sets that are depicted by the same graph are identical. That is, besides being extensional, sets are also *graph-extensional*. Slightly more formally, for sets a and b, let $a \equiv b$ if and only if there is an apg which is a picture of both a and b. Then, the graph-analogue of the principle of extensionality for sets, the Principle of Graph-Extensionality, can be formulated as follows:

$$\forall x \forall y (x \equiv y \rightarrow x = y).^{[17]}$$ (G-Ext)

As we pointed out in Chapter 1, it is widely believed that for a theory to be a theory of *sets*, sets have to be identical if they have the same members: it is usually thought to be part of our concept of set that the identity of sets is determined by their members. In a similar fashion, the graph conception – by taking sets to be what corresponds to graphs of the appropriate form – demands that the identity of sets be determined by the graphs they are depicted by. Clearly, however, the Principle of Graph-Extensionality is equivalent to AFA$_2$ (see Aczel 1988: 20, exercise 2.2). Again, we see that the uniqueness part of AFA is justified on the graph conception.

To this, one could object that all that can be said about equality between sets is that if two of them have the same members, they are identical. But what considerations can be offered in favour of this claim? One might try and appeal to the fact that, in the case of well-founded sets, as soon as the equality relation between the members of two sets has been fixed, the Axiom of Extensionality determines the equality conditions for the two sets: a straightforward application of transfinite induction on the membership relation then establishes that the equality relation between well-founded sets is uniquely determined (see Aczel 1988: 19).

But similarly to Aczel's unsuccessful argument for AFA, this argument does not work, since it is precisely well-foundedness that guarantees that fixing the equality relation between the members of the two sets suffices to determine their equality conditions. And indeed, as we saw when considering the graph shown in Figure 7.2, once we consider non-well-founded sets, there are cases in which the Axiom of Extensionality does not help to establish whether two sets are equal or not. In fact, Rieger himself subscribes to the idea that to deal with non-well-founded sets, we need some further criterion of equality besides Extensionality – we need to sharpen the criterion of identity for sets (see Section 1.6). For the letter of the Axiom of Extensionality only implies that sets should be equal when they have the same members, and if this is our only constraint, we end up with something like Boffa set theory, whereas Rieger's own view is that the set theory we should favour is Finsler-Aczel set theory.

---

[17] Incidentally, the relation $\equiv$ turns out to be equivalent to the largest bisimulation on the universe of sets V, when the notion of a graph figuring in the definition of bisimulation is widened so as to allow there to be a proper class of nodes. See Aczel 1988: 13 and 20–23 for details.

Alternatively, one might insist, more generally, that there is nothing more to sets than their extensional nature. One could then argue that this nature only demands that exact pictures be extensional or, following Rieger, that they be extensional and isomorphism-extensional. Either way, the upshot would be that we should not accept the Principle of Graph-Extensionality, since, the objection goes, sets do not have to obey it in order for their extensional nature to be respected.

This objection does not work either, however. For, as Rieger himself (2000: 245) seems to assume, the extensional nature of sets seems to consist in the fact that their membership structure should determine their identity. But, as I have argued, taking the membership structure of sets to decide questions of identity between them is enough to guarantee the truth of the Principle of Graph-Extensionality. It is worth pointing out, though, that the objection under consideration simply takes it for granted that sets have an extensional nature. Our strategy, on the other hand, has been to describe a conception of set which embodies the extensional nature of sets by being an articulation of the idea that sets are simply objects having membership structure. This is why all we claim to have shown is that the Principle of Graph-Extensionality is true on the graph conception, not that it is true *tout court*. Note, moreover, that the objection's assumption that there is nothing more to sets than their extensional nature – independently of whether this is the case on a particular conception of set – is also problematic. For there are conceptions of set on which sets do emerge as having some further feature besides their extensional nature. One such conception is the iterative conception, according to which sets are well-founded because they are the objects which can be obtained by iterated applications of the *set of* operation.

I conclude that Rieger's claim that if we admit non-well-founded sets at all, we ought to admit more than there are in the AFA universe is misguided: if we admit non-well-founded sets because we take sets to be the things that are depicted by graphs, then we should accept AFA. This also shows that the graph conception provides a reason for accepting the Axiom of Extensionality other than the fact that it is part of our concept of set that no two distinct sets can have the same members. For recall Aczel's key result that AFA is equivalent to the statement that an apg is an exact picture just in case it is strongly-extensional. A corollary of this result is that AFA implies that a graph is an exact picture only if it is extensional, since it is easy to see that there is a largest bisimulation between any two nodes having the same children (see Aczel 1988: 22, exercise 2.8.i). This corollary justifies Aczel's (1988: 23) assertion that AFA is a 'strengthening of extensionality', and shows that if the identity of sets is determined by the graphs they are depicted by, as the graph conception demands, it is also determined by the members they have.

Thus, the graph conception provides a prima facie reason for restricting attention to ZFA among the four non-well-founded set theories. However, the fact that AFA, and consequently Extensionality, are justified on the graph conception does not show that the same is true for the *other* axioms of ZFA. Our next task is therefore to examine the status of these axioms on the conception.

7.5 The Graph Conception and ZFA

Recall that Z_i is the theory whose axioms are those of ZFC minus Extensionality, Replacement and Choice. Boolos (1971; 1989) argued that all the Z_i axioms but none of the other ZFC axioms are true on the iterative conception. In support of this claim, he showed that the axioms of Z_i follow from the theory of stages, which, he claimed, captures (part of) the content of the iterative conception (see Section 2.2 for the axioms of ST, Boolos's most recent version of the theory of stages). The main purpose of this section is to show that similar considerations can be offered in support of the claim that, besides AFA and Extensionality, the axioms of Z_i^- (Z_i minus the Foundation Axiom) are also justified on the graph conception.

Recall that the axioms of Z_i^- divide rather neatly into two categories: axioms which make an outright existential assertion, of which the Axiom of Empty Set (if it is included) and the Axiom of Infinity are the only representatives; and axioms which make a *conditional* existential assertion, by stating that *if* there are certain sets, then there are also certain other sets. Our strategy will be as follows. We shall first offer informal considerations for the claim that, to the extent that they are true on the iterative conception, the axioms that make an outright existential assertion are true on the graph conception too. We shall then present an elementary theory of graphs from which the remaining axioms of Z_i^- follow. Thus, insofar as the theory can be taken to be formalizing part of the content of the graph conception, the axioms of Z_i^- which make a conditional existential assertion are true on the conception. We shall conclude with some remarks on the status of the remaining axioms of ZFA. Note that although AFA is justified on the graph conception, nothing we shall say, except for some brief remarks on the Axiom of Replacement, will hinge on this fact: if the axioms of Z_i^- are true on the graph conception, they are so independently of the fact that AFA is justified on the conception.

Let us start from the Axiom of Empty Set, which is easily seen to be true on the graph conception. For if any apg is a graph in good standing, then *surely* the graph consisting of just one childless node is. So if sets are what is depicted by an arbitrary graph, the set depicted by this graph exists, and, of course, this set is the empty set.

What about the Axiom of Infinity? The axiom comes in a variety of alternative formulations, not always interderivable in first-order logic (see Section 2.1, fn. 4). Here, we shall focus on the version of the axiom described in Section 2.1, that is

$$\exists y(\emptyset \in y \wedge \forall x(x \in y \rightarrow x \cup \{x\} \in y)),$$

but what we shall say carries over to the other formulations. This version of the axiom delivers (in the presence of the Separation Schema) the existence of the infinite set

$$\mathbf{N} = \{\emptyset, \{\emptyset\}, \{\emptyset, \{\emptyset\}\}, \{\emptyset, \{\emptyset\}, \{\emptyset, \{\emptyset, \{\emptyset\}\}\}\}, \ldots\},$$

which is the smallest inductive set. And the existence of this set, in turn, guarantees the truth of the Axiom of Infinity. Thus, if we can offer reasons for thinking that there is a graph depicting this set, we will have thereby offered reasons for thinking that the axiom is true on the graph conception.

The graph in question can be constructed as follows. We start with the one-node childless graph, which we know to be a legitimate graph. Each time we are given or have constructed a graph $\mathcal{G}$, we construct a new graph $\mathcal{H}$ by drawing a new edge from the top node of $\mathcal{G}$ to a node n which, in turn, is the top node of a subgraph of $\mathcal{H}$ isomorphic to $\mathcal{G}$. More intuitively, each new graph in the series can be constructed by simply taking two isomorphic copies of the given graph $\mathcal{G}$ and drawing an edge from the top node of one of them to the top node of the other. We repeat this procedure infinitely many times; by the end of it, we will have obtained a graph depicting **N**. The beginning of the construction is illustrated in Figure 7.7.

Is this construction legitimate on the graph conception? The crucial step is the possibility of performing an operation, such as that of forming copies of a given graph and that of drawing new edges, infinitely many times. But the possibility of doing so seems inherent to the graph conception. On this conception, *any* apg is a graph in good standing, which is what is meant to lead us to accepting, for instance, the existence of trees that have an infinite descending path, such as the one in Figure 7.4. And graphs of this kind exist only if we allow for the possibility of drawing edges infinitely many times. Thus, the same considerations that lead to accepting the existence of trees that have an infinite descending path lead to accepting the existence of a graph with infinitely many edges emanating from its top node. Clearly, this is not to suggest that no doubts can be raised *in general* about the possibility of performing an operation infinitely many times. Rather, the claim is that any such doubts have *already* been set aside in order to accept the existence of, e.g., trees with an infinite descending path.

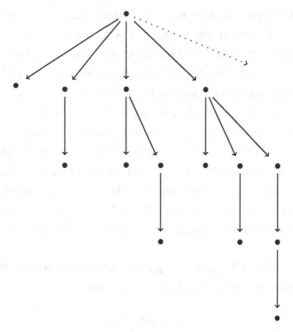

Figure 7.7 Constructing the graph depicting the set **N**.

One might insist that it is precisely these doubts that prevent the Axiom of Infinity from being justified on the graph conception. If that is true, however, these doubts will also undermine, *contra* the received view on the matter,[18] the usual arguments that the axiom is justified on the iterative conception. For these arguments too rely, one way or another, on the possibility of performing an operation infinitely many times, namely the *set of* operation. For instance, Boolos's stage theory delivers Infinity because it includes an axiom asserting that there exists an infinite stage in the iterative process of set formation. Thus, the graph conception's justification of the Axiom of Infinity and the one standardly offered by the iterative conception seem to stand or fall together.

Having presented some considerations for thinking that the Axioms of Empty Set and Infinity are true on the graph conception, we now need to turn to the theory of graphs intended to show that the remaining axioms of Z_i^- follow from the graph conception. More precisely, the theory we shall present is a theory of *trees*, modelled upon Boolos's (1989) stage theory ST. The restriction to trees will make it easier to provide a theory whose axioms can be seen to be true on the graph

---

[18] Besides Boolos 1971 and Boolos 1989, see, e.g., Potter 2004: 68–72 and Paseau 2007: 6.

conception and to imply the axioms of Z_i^- which make a conditional existential assertion. On the other hand, the restriction is harmless since, as we pointed out above, every graph depicting a certain set can be unfolded into a tree depicting the same set. As we shall shortly see, this fact will turn out to be crucial when discussing one of the axioms of our theory of trees.

The theory, to which we shall refer as T, is cast in a two-sorted first-order language $\mathcal{L}$ with variables $x, y, z, \ldots$ ranging over sets and variables $g, h, i, \ldots$ ranging over trees. We have three two-place predicates: a tree-tree predicate $\unlhd$, which may be read 'is a subtree of'; a set-tree predicate D, which may be read 'is depicted by', and a set-set predicate $\in$, to be read in the usual way. Following Boolos, we abbreviate '$\exists h(h \unlhd g \wedge yDh)$' as ySg, one possible reading of which is 'y is depicted by a subtree of g'; and, as usual, we abbreviate '$\forall z(z \in x \rightarrow z \in y)$' as $x \subseteq y$.

Now for the axioms of T and what can be said on their behalf. First, we have an axiom stating that every set is depicted by some tree:

$$All \quad \forall x \exists g\, x\, Dg.$$

It is clearly the case that on the graph conception every set is depicted by some graph: if sets are what is depicted by an arbitrary apg, then for any set a there must be an apg depicting a. By itself, this would suffice to make *All* true if the variable g were taken to range over apgs; but since every apg depicting a certain set can be unfolded into a tree depicting the same set, it also shows that *All* is true on the graph conception when g is taken, as it should, as ranging over trees.

Next, we have two axioms specifying the structural features of $\unlhd$:

$$Tra \quad \forall g \forall h \forall i (g \unlhd h \wedge h \unlhd i \rightarrow g \unlhd i).$$
$$Dir \quad \forall g \forall h \exists i (g \unlhd i \wedge h \unlhd i).$$

Tra says that $\unlhd$ is transitive. To begin with, notice that if *Tra* holds when $\unlhd$ is read as 'is a subgraph of', then it obviously holds when it is read as 'is a subtree of'. But it is easy to see that *Tra* holds when $\unlhd$ is read as 'is a subgraph of': if a graph $\mathcal{G}$ consists of nodes taken from a graph $\mathcal{H}$ and all edges of $\mathcal{H}$ between these nodes, and $\mathcal{H}$ consists of nodes taken from a graph $\mathcal{I}$ and all edges of $\mathcal{I}$ between these nodes, then $\mathcal{G}$ will consist of nodes taken from $\mathcal{I}$ and all edges of $\mathcal{I}$ between these nodes. This is reflected in the fact – on which, however, we shall not officially rely, given what we said in Section 7.3 – that the transitivity of the relation *is a subgraph of* is an elementary theorem of graph theory, since it is an immediate consequence of the set-theoretic definition of subgraph (see Tutte 2001: 10).

Dir says that $\unlhd$ is directed. To convince yourself of its truth on the graph conception, just reflect on the fact that given any two trees $\mathcal{G}$ and $\mathcal{H}$ with, respectively, roots n_0 and n_1, it is always possible to form a tree $\mathcal{I}$ which has, in addition to all the nodes and edges of $\mathcal{G}$ and $\mathcal{H}$, a node n_2 as root and edges $n_2 \longrightarrow n_0$ and $n_2 \longrightarrow n_1$. $\mathcal{I}$ will then be the required tree having $\mathcal{G}$ and $\mathcal{H}$ as subtrees.

The third group of axioms tells us how the membership and subsethood constituencies of the set depicted by a tree are reflected in the internal structure of the tree:

$$Mem \ \forall x \forall g (x D g \rightarrow \forall y (y \in x \rightarrow y S g)).$$

$$Sub \ \forall x \forall g (x D g \rightarrow \forall y (y \subseteq x \rightarrow y S g)).$$

Mem states that if a set a is depicted by a tree $\mathcal{G}$, then each of its members is depicted by a subtree of $\mathcal{G}$. To see that this principle is true on the graph conception, it suffices to note that each $b \in a$ is depicted by the subtree of $\mathcal{G}$ below the node representing b. What is crucial here is that, since we are dealing with trees, there is only one path from $\mathcal{G}$'s root to any of the nodes belonging to the subtree below the node representing b. Hence, none of these nodes is connected to any node of any subtree below nodes representing any other member of a.

Note that *Mem* corresponds to the left-to-right direction of the axiom *When* of the stage theory ST. The reason why the corresponding right-to-left direction is not included in our theory of trees is that it is obviously false on the graph conception. Consider, for instance, the set $\{\varnothing, \{\varnothing\}\}$. It is easy to see that each of its members is depicted by a subtree of the tree in Figure 7.8. Obviously enough, however, the set depicted by this tree is $\{\{\varnothing\}\}$. Far from being a problem, this is good news, since the right-to-left direction of *When* delivers Foundation in the presence of the other axioms of T (Boolos 1989: 95–96). *When*, however, is also used by Boolos

Figure 7.8 Graph depicting the set $\{\{\varnothing\}\}$.

to derive Powerset. For this reason, we have had to add a new axiom to our theory, namely *Sub*.

This states that if a set a is depicted by a tree $\mathcal{G}$, then each of its subsets is depicted by a subtree of $\mathcal{G}$. What we need to check is that, no matter what a looks like, any $b \subseteq a$ will be depicted by a subtree of $\mathcal{G}$. There are three cases to consider. If b is the empty set, then it is obviously depicted by a subtree of $\mathcal{G}$, namely the one consisting solely of $\mathcal{G}$'s root. Similarly, if b is a itself, the subtree of $\mathcal{G}$ depicting b will be $\mathcal{G}$ itself. There remains the case where b is a proper subset of a other than the empty set. We know that, for each member of a, there is an edge from $\mathcal{G}$'s root to the node representing the member of a in question. And, as in the case of *Mem*, we also know that no node of the subtree below the node representing the member of a in question will be connected to any node of any subtree below nodes representing any other member of a. Thus, we can obtain the subtree of $\mathcal{G}$ depicting b by, as it were, removing the edges of $\mathcal{G}$ connected to nodes representing the members of a which are not in b (together with the subtrees below the nodes representing these members).

The last group of axioms consists of the specification axioms, i.e. all instances of the schema

$$Spec \quad \exists g \forall y (\varphi(y) \rightarrow ySg) \rightarrow \exists x \forall y (y \in x \leftrightarrow \varphi(y)),$$

where $\varphi(y)$ is any formula of $\mathcal{L}$ containing no free occurrences of x.

Spec says that if all sets y satisfying a certain condition φ are depicted by subtrees of a tree $\mathcal{G}$, they form a set. To see that the axiom holds on the graph conception, we construct a new tree $\mathcal{H}$ as follows. We take a node n, which will be the root of our tree, and, for each y, we draw an edge from this node to the root of the subtree of $\mathcal{G}$ depicting y. Given the guiding principle of the graph conception that any apg is a graph in good standing, $\mathcal{H}$ is a well-formed graph; but it is easy to see that $\mathcal{H}$ will depict the set of all ys satisfying φ.

Spec seeks to make sure that any set depicted by a tree exists, and it does so by requiring that whenever all sets satisfying a certain condition are depicted by subtrees of a certain tree, they form a set. Thus, the axiom attempts to capture (for the case of trees) the thought that on the graph conception sets are what is depicted by an *arbitrary* graph.

This completes the presentation and discussion of the axioms of T. As anticipated, we have the following result: (Except for the case of Powerset, the proof is adapted from Boolos's paper.)

Theorem 7.5.1. T *implies the Axioms of Unordered Pairs, Union, Powerset and Separation.*

Proof. By *All*, for some g and h, zDg and wDh. By *Dir*, for some i, $g \trianglelefteq i$ and $h \trianglelefteq i$. Thus, zSi and wSi. So we have shown that $\exists g \forall y ((y = z \lor y = w) \rightarrow ySg)$. By the relevant instance of *Spec*, we get, by *modus ponens*, the Axiom of Unordered Pairs.

By *All*, for some g, zDg. Thus, if $w \in z$, by *Mem*, for some h, $h \trianglelefteq g$ and wDh. If $y \in w$, then, by *Mem* again, for some i, $i \trianglelefteq h$ and yDi. But from $h \trianglelefteq g$ and $i \trianglelefteq h$, by *Tra*, we have $i \trianglelefteq g$, and so ySg. Thus, we have established that $\exists w (y \in w \land w \in z) \rightarrow ySg$. But $(\exists w (y \in w \land w \in z) \rightarrow ySg) \rightarrow \exists x \forall y (y \in x \leftrightarrow \exists w (y \in w \land w \in z)$ is an instance of *Spec*. By *modus ponens*, we obtain Union.

By *All*, for some g, zDg. By *Sub*, $\forall y (y \subseteq z \rightarrow ySg)$. But $\forall y (y \subseteq z \rightarrow ySg) \rightarrow \exists x \forall y (y \in x \leftrightarrow y \subseteq z)$ is an instance of *Spec*; so, by *modus ponens*, we get Powerset.

By *All*, for some g, zDg. By *Mem*, $y \in z \rightarrow ySg$. *A fortiori*, $y \in z \land \varphi(y) \rightarrow ySg$. But $\forall y (y \in z \land \varphi(y) \rightarrow ySg) \rightarrow \exists x \forall y (y \in x \leftrightarrow \varphi(y))$ is an instance of *Spec*; so, by *modus ponens*, we obtain Separation.   □

Thus, insofar as the considerations offered above show that the axioms of T are true on the graph conception, the axioms of Z_i^- which make a conditional existential assertion are true on this conception too. This concludes our case that the axioms of Z_i^- plus AFA and Extensionality are justified on the graph conception.[19]

But what about the remaining axioms of ZFA, *viz.* Replacement and Choice? As we have seen, the question whether these two axioms are justified on the iterative conception is controversial, and similar issues arise in the current case. Here, we will limit ourselves to providing reasons for thinking that the graph conception does just as well (or just as badly) as the iterative conception at justifying these axioms.

Let us consider the Axiom of Choice first. Recall that the standard argument that AC is true on the iterative conception says that if each level of the hierarchy contains all subsets of the previous levels (and not only those defined by the Separation Axiom), then it will *a fortiori* contain the choice sets (see Section 2.1).

The complaint Boolos (1971: 28–29; 1989: 96–97) levelled against this argument is, essentially, that it just *assumes* that the choice sets are among the subsets of a given set. For it is true that each level of the hierarchy contains all subsets of sets

---

[19] Baltag (1999: 484–485) has claimed that a conception of set lying behind a non-well-founded set theory must fail to sanction Separation: once the iterative conception is no longer available, the axiom can only be justified by placing explicit size restrictions on admissible sets. Our findings suggest otherwise: the graph conception takes sets to be what is depicted by an *arbitrary* graph, and hence seems to impose no explicit smallness condition on sets; but it does validate the Separation Axiom, and indeed many of the ZFC⁻ axioms.

occurring at previous levels. And it is also true that all members of members of that level occur at previous levels. But this only gives us Choice if we assume that the choice sets for a set a are among the subsets of $\bigcup a$ (which is uncontroversially a set). And to assume this, Boolos claimed, is to assume precisely what is at stake in discussions over the truth of the Axiom of Choice.

The situation is analogous in the case of the graph conception. For, as the derivation of the Axiom of Union from T shows, if a is a set, then $\bigcup a$ is a set too, and is therefore depicted by a tree $\mathcal{G}$. But, one might argue, *all* subsets of $\bigcup a$ are depicted by a subtree of $\mathcal{G}$, including the choice sets for a. And adding an axiom to this effect to our theory of trees would immediately deliver Choice. However, following Boolos, one might complain that this argument begs the question, since, in effect, it assumes that the choice sets are among the subsets of $\bigcup A$, which is exactly what a sceptic about Choice would be sceptical about.

Hence, the standard argument that Choice is true on the iterative conception can be easily turned into an argument that the axiom is true on the graph conception, and similarly for Boolos's objection. Moreover, if there is any reason to think that AC is true on the iterative conception because this is best understood as a combinatorial conception, this would also apply to the graph conception, since this conception too is naturally regarded as a combinatorial conception. The upshot is that there is at present no reason to think that the graph conception does worse than the iterative conception with respect to the Axiom of Choice.

Let us now turn to the Axiom of Replacement. The axiom, recall, states that the image of a set under a first-order specifiable function is also a set. Thus, in terms of the graph conception, the truth of the axiom demands that each time we are given a definable function f and a graph depicting a set a, there is a graph depicting the set whose elements are precisely the $f(x)$s for $x \in a$. Whether any principle to the effect that such a graph always exists can be justified on the basis of the graph conception alone does appear to be doubtful. But even admitting that, it does not follow that the iterative conception is better off than the graph conception with respect to the Axiom of Replacement. For it is very controversial whether each instance of the axiom is justified on the iterative conception unless this is taken to incorporate the idea that the *set of* operation is to be iterated as far as possible (see Section 3.6), and a similar idea could be incorporated into the graph conception.

However, there is a point in this connection which I think deserves mention. One central application of the Axiom of Replacement in standard, well-founded set theory is in the theory of ordinals. In particular, if we define ordinals in the standard way – that is, as transitive sets well-ordered by membership – we need Replacement to prove the Mirimanoff-von Neumann result that every well-ordered set is isomorphic to an ordinal. The situation is different when the Axiom of Foundation is replaced with AFA (which, if the arguments offered in Section 7.4 are correct, is justified on the graph conception). For the Mirimanoff-von Neumann result is a special case of Mostowski's Collapsing Lemma, whose graph-theoretic version is simply the statement – first encountered in Section 7.1 – that every well-founded graph has a unique decoration. But AFA tells us that every graph – and hence every *well-founded* graph – has a unique decoration. Thus, in its presence we can prove the graph-theoretic version of the Collapsing Lemma and develop the standard theory of ordinals without Replacement.[20]

7.6 The No New Isomorphism Types Objection

We now turn to four objections which have been raised against non-well-founded set theories and purport to provide us with reasons for favouring an iterative set theory over a non-well-founded set theory.

The first objection is that we should favour an iterative set theory because the restriction to well-founded sets – although not necessary to avoid inconsistency – is not a real restriction from the point of view of the mathematician. Let me explain. Recall that in Chapter 2, I mentioned the fact that the reason why the restriction to pure sets is not really regarded as a restriction by mathematicians is that it does not seem to prevent the existence of any new isomorphism type. A similar explanation can be given concerning the restriction to well-founded sets in the presence of the axioms of ZFC⁻: there are no isomorphism types that are realized in the non-well-founded universe of, say, ZFA that are not already realized in the well-founded universe of ZFC. So, for instance, although AFA forces the existence of non-well-founded sets such as $Ʊ = \{Ʊ\}$, the isomorphism type of this set is already realized in the well-founded universe, albeit not by the membership relation but by other relations, such as edges.[21] What this means is that the structures which are given to us by ZFA are already available in ZFC and can therefore be studied by mathematicians without having to resort to different set theories.

This suggests the following reason for restricting attention to well-founded sets: the restriction, whilst making things technically smoother, does not prevent the existence of any new isomorphism type. However, as an *objection* to the adoption

---

[20] These remarks should not be taken as suggesting that no theory of ordinals can be developed in well-founded set theory unless Replacement is available. For if we replace ZFC's Axiom of Foundation with axiom (ρ) and use the Scott-Tarski definition of an ordinal (see Section 3.2), we can develop the theory of ordinals without Replacement. See, e.g., Potter 2004: 175–190 for details.

[21] The point was first made by McLarty (1993), and has been reiterated and widely used by Maddy (1996; 1997).

of a non-well-founded set theory of the kind under consideration, this is not very convincing. For a start, as we pointed out in Section 2.1 for the case of individuals, the fact that it is technically smoother to restrict attention to well-founded sets should not lead us to draw too drastic conclusions from it. We certainly do not want to conclude that there are no sets with individuals simply on the grounds that it is technically smoother to restrict attention to individuals; similarly, it would be too quick to conclude that there are no non-well-founded sets on the grounds that the restriction to well-founded sets is technically convenient.

But even setting this aside, the objection does not work, since it is far from clear that the restriction to well-founded sets really makes things technically smoother. On the one hand, it is easy to see that the cumulative hierarchy is an inner model of the non-well-founded universe of ZFA. Thus, for certain purposes, one can restrict attention to the sets that appear at some level or another of the cumulative hierarchy whilst holding on to the idea that the hierarchy does not exhaust all sets. The point, here, is analogous to one made by Maddy (1997: Ch. 6) in her well-known discussion of the Axiom of Constructibility. A well-known result of Scott (1961) establishes that the existence of a measurable cardinal implies the existence of non-constructible sets. Nonetheless, Maddy argues, someone who accepts the existence of measurable cardinals can, for certain purposes, restrict attention to inner models in which **V=L** holds.

On the other hand, the fact that in a non-well-founded set theory certain structures are realized by the membership relation might make things easier in certain applications (as in, for instance, Barwise and Etchemendy 1987; Barwise and Moss 1996). Hence, we can conclude that 'this objection to the adoption of AFA as an axiom only makes sense if one is already committed to the axiom of Foundation' (Antonelli 1999: 158).

7.7 The No Intuitive Model Objection

The second objection – raised by, e.g., Maddy (1997) and closely related to the worry that there is no conception of set lying behind a non-well-founded set theory – is that there is no intuitive conception of the AFA universe. On the other hand, the concern continues, we have such a conception of the ZFC universe in the cumulative hierarchy, which is therefore a reason for preferring an iterative set theory over a non-well-founded one.

The problem with the objection is that there *is* a hierarchical construction of the AFA universe, which provides us with a picture of what this universe looks like. The construction proceeds in a very similar fashion to the more familiar hierarchical construction which gives rise to the cumulative hierarchy. As in the case of the iterative conception, but restricting attention to pure sets, at the beginning we have the empty set. At any finite stage, we form all possible collections of items formed at earlier stages and all collections depicted by apgs whose nodes are taken from the apgs depicting all possible collections of items formed at earlier stages.[22] After the finite stages come the infinite ones, corresponding to the ordinals ω, $\omega + 1$, $\omega + 2$, $\omega + \omega$, etc. At limit stages, we form all arbitrary collections of items formed at earlier stages. The result is, similarly to the case of the iterative conception, a picture of the set-theoretic universe as a cumulative hierarchy divided into levels. The picture can be captured by the following characterization. Let α be an ordinal. Then, the levels of what we may call the *graph-theoretic hierarchy*, **G**, are organized as follows:

$$G_0 = \emptyset;$$
$$G_{\alpha+1} = \mathcal{C}(G_\alpha);$$
$$G_\lambda = \bigcup_{\alpha < \lambda} G_\alpha \text{ if } \lambda \text{ is a limit ordinal,}$$

where $\mathcal{C}$ is the operation which, given a set a, produces the set of all subsets of a and all sets depicted by apgs whose nodes are taken from apgs depicting subsets of a.[23]

This characterization of the set-theoretic universe is implicit in Aczel's proof that, if ZFC is consistent, then so is ZFA (see Aczel 1988: 33–37). For Aczel's result, in effect, tells us that, given a model of ZFC_2, we can construct a model of ZFA_2, the second-order version of ZFA.[24] This theory, then, can be shown to be quasi-categorical in the same way as ZFC_2: just as the parts of two models of ZFC_2 containing sets of less than or equal to a given rank are isomorphic, the parts of two models of ZFA_2 containing sets whose transitive closure is less than or equal to some given cardinality are isomorphic too. This means that we have a hierarchical construction of the non-well-founded universe of ZFA, which determines a natural ordering on the universe of sets, based on their complexity, as measured in terms of cardinality of their transitive closure.

It should be stressed that the iterative construction just described is not meant to provide us with a conception of what sets *are* on the graph conception – they

---

[22] This is essentially the construction described in Forster 2008: 108.

[23] For a related construction, see Holmes 2009.

[24] What Aczel's original result shows is that, given a full model of ZFC, we can construct a full model of ZFA which is unique up to isomorphism – where a model is said to be *full* if all its subsets occur as sets of children of some unique node. What this means, roughly, is that a full model of ZFC will be a model of ZFC_2.

Figure 7.9 Graph depicting one of the sets occurring at level G_1.

are the things depicted by graphs, as the graph conception maintains. What the iterative construction is intended to provide us with, to stress, is a picture of what the *set-theoretic universe* looks like on the graph conception. The importance of this fact becomes clear once we consider what happens, in the construction of the graph-theoretic hierarchy, when we move from a level to its successor.

Consider, for instance, level 0. Here, as in the hierarchy of well-founded sets, we have the empty set. To reach the next level, recall, we have to form all possible collections of items formed at earlier stages and all collections depicted by apgs whose nodes are taken from the apgs depicting the collections just formed. This means that level 1 will include the set depicted by the graph in Figure 7.9, that is the set $\mho = \{\mho\}$. This brings to light one fact that is easy to overlook when considering the construction which gives rise to the graph-theoretic hierarchy, namely that the connection between the sets occurring at one level of it and the sets occurring at the next level must be explicated in terms of the graphs these sets are depicted by, and cannot be explained by mentioning only their members. For consider the set $\mho$. The membership constituency of this set has no connection whatsoever with the empty set: this set has $\mho$ as a member, which in turn has $\mho$ as a member, which, again, has $\mho$ as a member, and so on. The connection between this set and the empty set is rather explained by the fact that the graph which depicts it is obtained by drawing a loop starting from the one-node graph which depicts precisely the empty set.

As a result, it is unlikely that the construction can be described in a rough, but informative way by using a primitive of which we have an independent grasp as in the case of the iterative conception. Let me explain. In the case of the iterative conception, the operation of going from one level of the cumulative hierarchy of well-founded sets to its successor is given by the powerset operation, which is a formal rendering of the notion of *set of*. This means that we can describe the iterative conception by saying (for the case of pure sets) that the sets are the empty set and the set containing the empty set, sets of those, sets of *those*, and so on, and then there is an infinite set, and sets of its elements, sets of those, sets of *those*, and so on. But can we convey the iteration of graph conception in a similar way? The fact that there is no connection in terms of membership constituency between a level and its successor in the graph-theoretic hierarchy means that the rough description has to proceed by mentioning relevant features of the graphs concerned.

Two possibilities suggest themselves. One possibility would be to say (again for the case of pure sets) that the sets are the set depicted by the one-node graph, and then sets depicted by graphs of that set, and then sets depicted by graphs of those, and so on. The problem with this characterization is that it is far from clear that we have a grasp of the *graph of* operation. What we have a grasp of, perhaps, is the notion of all the apgs which can be formed given certain nodes. This brings us to the second possibility, which consists in saying that the sets are the empty set, the set containing the empty set plus all the sets which are depicted by graphs constructed using nodes which depict this new set, sets of those plus all the sets which are depicted by graphs constructed using nodes which depict these new sets, and so on, and then there is an infinite set, and sets of its elements plus all the sets which are depicted by graphs constructed using nodes which depict these new sets, and so on. But it seems unlikely that a conception of sets as what is obtained by iterating the operation *set of, plus all sets which are depicted by graphs constructed using these nodes* would be deemed to be natural. But, and this is the crucial point, the hierarchical construction of the AFA universe described in this section is only intended to provide us with a picture of this universe, not with a conception of what sets are. The latter task is performed by the graph conception of set itself, which tells us that sets are what is depicted by an arbitrary graph.

7.8 The No Place for *Urelemente* Objection

In this chapter, we have restricted attention to *pure* non-well-founded set theory, in keeping with the policy adopted throughout the book. Still, if non-well-founded set theory is to serve in *applications*, it should be possible to formulate it in such a way that *Urelemente* are countenanced, and, importantly, we should be able to justify the resulting theory in a similar way as we did in the case of its pure counterpart. The third objection to non-well-founded set theory is that this cannot be done.

Now recall that a decoration of a graph is an assignment of elements of the set-theoretic universe to each node of the graph such that the elements of the set assigned to the node are the sets assigned to the children of the node. This definition has the consequence that a childness node must be assigned the empty set and, therefore, that graphs can only depict pure sets. As mentioned, however, our restriction to pure sets has only been a matter of convenience, and one can obtain impure versions of non-well-founded set theory using techniques developed in Barwise and Etchemendy 1987: 39–40 and Barwise and Moss 1996: 125–130.

Barwise and Etchemendy's strategy, in particular, fits very well with the graph conception of set. On this strategy, we modify the notion of a decoration so that graphs can depict impure sets: as before, we regard childless nodes as representing entities which do not have members; but these entities can now include objects other than the empty set. Formally, one considers *tagged* graphs, that is graphs whose childless nodes have been assigned either the empty set or an individual by a *tagging*. A decoration of a tagged graph will then assign to the childless nodes what is assigned to them by the tagging and to each other node n elements of the set-theoretic universe such that the elements of the set assigned to n are the sets or individuals assigned to the children of n.

Thus, even if we countenance impure sets, we can take sets to be the things depicted by graphs, as the graph conception maintains. Admittedly, however, the resulting version of the graph conception is less natural than the original version. For we are now taking sets to be what is depicted by *tagged* graphs, and so what set is being depicted is determined not only by the graph-theoretic structure of the graph (i.e. its nodes and edges) but also by the tagging of its childless nodes.

To appreciate this point further, recall the possible motivation for the graph conception we gave in Section 7.3. According to that motivation, we are led to taking sets to be things depicted by graphs by thinking of sets as simply objects having membership structure. But this is precisely what we are no longer doing when we take what a graph depicts to be partly determined by its tagging. For in that case, we are taking what a set *is* to be determined by the *Urelemente* it has as well as its membership structure.

Thus, it seems that whilst the graph conception can be modified so as to encompass sets with *Urelemente*, it seems a more natural conception when our focus is on pure sets.

7.9 The No Autonomy Objection

The fourth objection is that a non-well-founded set theory cannot aspire to be a foundation for the whole of mathematics because it lacks the required *autonomy*. The issue of autonomous foundations has been discussed in connection with the issue of whether category theory or homotopy type theory can serve as a foundation for mathematics which is suitably autonomous from set-theoretic foundations (for the case of category theory, see Feferman 1977; 2012, for the case of homotopy type theory, see Ladyman and Presnell 2018). To determine whether non-well-founded set theory can be regarded as adequately autonomous, it will be helpful to follow Linnebo and Pettigrew (2011) in distinguishing between three respects in which a proposed foundation of mathematics can be autonomous, namely *logical*, *conceptual* and *justificatory*. In particular, let T_1 and T_2 be theories. Then Linnebo and Pettigrew characterize the three kinds of autonomy as follows:

- T_1 has *logical autonomy* with respect to T_2 if it is possible to formulate T_1 without appealing to notions that belong to T_2.
- T_1 has *conceptual autonomy* with respect to T_2 if it is possible to understand T_1 without first understanding notions that belong to T_2.
- T_1 has *justificatory autonomy* with respect to T_2 if it is possible to motivate and justify the claims of T_1 without appealing to T_2, or to justifications that belong to T_2.

At first sight, these characterizations look implausible. For instance, according to the letter of the characterization of logical autonomy, it would seem to follow that no theory formulated in $\mathcal{L}_\in$ is logically autonomous from any other theory formulated in that language. However, as Linnebo and Pettigrew's subsequent discussion makes clear, the characterization should be understood so that the logical autonomy of T_1 with respect to T_2 is only compromised if one must appeal to notions belonging *specifically* to T_2 in order to formulate T_1. Similar considerations apply to the other characterizations of autonomy.

Now non-well-founded set theories do have logical autonomy with respect to set theories based on the iterative conception such as Z^+. For it is true that non-well-founded set theories are formulated in the language of $\mathcal{L}_\in$, and membership is a notion that belongs to well-founded set theories. But membership is not a notion that belongs *specifically* to well-founded set theory: non-well-founded set theory, if it is to provide an alternative foundation for mathematics at all, has an equal claim to membership.

Similarly, non-well-founded set theories are conceptually autonomous with respect to well-founded set theories. One does not need to understand notions belonging specifically to well-founded set theory in order to understand non-well-founded set theory.

But does non-well-founded set theory enjoy justificatory autonomy? As we have argued in this chapter, non-well-founded set theory is best seen as being justified on the basis of the graph conception of set, which sanctions most of the axioms of ZFA. We already stressed in Section 7.3 that if the graph conception is not to rest on a prior notion of set, thereby compromising its ability to serve as a foundation for set theory, it must be understood as talking about graphs not as set-theoretic constructions, but as entities that we can understand in more basic, primitive terms, without the mediation of set theory. This ensures that non-well-founded set theory is justificatory autonomous from set theories based on the iterative conception of set.

On the other hand, it also means that, ultimately, non-well-founded set theory must be justified by appealing to considerations that come from the theory of graphs. Thus, in this sense, non-well-founded set theory is *not* justificatory autonomous from graph theory.

But why should a foundation be required to have justificatory autonomy? Linnebo and Pettigrew, as well as subsequent discussions (e.g., Ernst 2017), do not go into detail here, but I think the criterion can be motivated by considering the view that a foundation of mathematics should provide the ultimate arbiter for questions of truth and existence in mathematics (see Maddy 1997; 2007; 2011). According to this view, one purpose of foundations is that should a dispute arise about the existence of a certain mathematical object, it should be possible to solve the dispute by establishing whether the object exists in the set-theoretic arena. But if the answer to the question of whether such an object exists in the set-theoretic arena depends, in turn, on the answer to questions from some other mathematical area, then set theory will *not* be the *ultimate* arbiter of truth and existence in mathematics.

To make things more concrete and apply the discussion to the case at hand, consider the case of two group theorists having a dispute as to whether a certain group exists. According to the view of a foundation as ultimate arbiter of truth and existence, it should be possible to settle their dispute by looking at whether the relevant group can be found in the set-theoretic arena. However, whether the latter is the case will, in turn, depend on whether we believe certain principles about graphs to hold in the first place, which we can then use to justify the axioms of non-well-founded set theory. Thus, it seems that set theory would have to forfeit its role as an ultimate arbiter of truth and existence.

Even worse, if we consider a dispute between two *graph* theorists about the existence of a certain graph, set theory's foundation role would require that it should be possible to settle the dispute by determining whether (a proxy for) the graph exists in the set-theoretic arena. But whether this is the case might depend on principles of graph existence about which the graph theorists might disagree in the first place. To put the point in a different way, if a foundational theory is to be an ultimate arbiter over questions of truth and existence in mathematics, it should be a suitably *neutral* arbiter. But if questions of truth and existence in set theory depend, in turn, on questions of truth and existence in graph theory, set theory will not be able to act as a neutral arbiter over questions of truth and existence in graph theory.

7.10 Conclusion

It is natural to make use of graphs to depict sets. And, once we do that, it is very natural to take sets to *be* what is depicted by graphs. Thus, the graph conception naturally emerges as a candidate to be the conception of set embodied by a set theory centred around the idea of a set being depicted by a graph. In fact, the graph conception turns out to be a rather *successful* candidate. For, I have argued, if sets are what is depicted by an arbitrary graph, then most of the axioms of ZFA are justified.

The graph conception, moreover, appears to be consistent, and the theory behind which it lies is consistent just in case ZFC is. Hence, the graph conception refutes Boolos's (1989: 90) claim that the iterative conception is, 'the only natural and (apparently) consistent conception of set we have'.

It does not follow, of course, that the axioms of non-well-founded set theories are justified *tout court*. We investigated this issue by considering four objections to the graph conception and non-well-founded set theory. The standard objection against non-well-founded set theories that they do not introduce new isomorphism types just assumes the Axiom of Foundation as a starting point. Moreover, non-well-founded set theory does have an intuitive model in the graph-theoretic hierarchy. However, the graph conception appears to be less natural once we consider it a conception of impure as well as pure sets. More importantly, non-well-founded set theory fails to provide a foundation for mathematics autonomous from graph theory. I conclude that, if set theory is to provide a foundation for mathematics in the sense articulated by Maddy and endorsed by many a set theorist, the iterative conception fares better than the graph conception.

8

Concluding Remarks

We have encountered a number of conceptions of set, each with its own peculiarities. It might therefore be useful to summarize their commonalities and differences.

An important theme has been the diagnosis of the paradoxes as emerging from the fact that, on the naïve conception, we have both indefinite extensibility and universality of the set concept. The iterative conception, the limitation of size conception, and the graph conception reject universality. The stratified conception, by contrast, rejects indefinite extensibility. The *spirit* of the definite conception is to reject universality, like the limitation of size conception, but its status here needs further investigation.

Another distinction which has featured prominently in the book is that between logical and combinatorial conceptions of collection. The iterative conception and the graph conception are rather naturally thought of as combinatorial conceptions, whereas the limitation of size conception, the definite conception and the stratified conception are best seen as logical conceptions of set. The naïve conception is the paradigmatic example of a logical conception, although we saw that the appeal of the Naïve Comprehension Schema – the distinguishing mark of the conception – may be due to the conflation between the combinatorial conception of iterative set and the logical conception of objectified property.

Finally, we have well-foundedness, a feature of iterative sets which the sets as described by any of the other conceptions need not have. Table 8.1 summarizes these findings.

This book can be read as an extended argument for the iterative conception of set and hence for any axioms which are justified on this conception. The question of what axioms those are deserves further exploration. Consider, for instance, large large cardinal axioms (see fn. 30). Are these axioms justified on the iterative conception of set? And what about axioms that purport to capture the idea that the hierarchy is as *wide* as possible, such as forcing axioms or the Inner Model Hypothesis (see Incurvati 2017)? The method of inference to the best conception promises to help answer these questions too. In particular, various ways of sharpening further the iterative conception seem available – for instance, as embodying some maximality assumption concerning the height or width of the hierarchy – and the method of inference to the best conception tells us that, ultimately, the conception to be preferred will be that which fares better than its rivals with respect to the desiderata we have been discussing.

This book is compatible with a moderate form of pluralism about conceptions of set, according to which, depending on the goals one has, different conceptions of set might be preferable. The book is also compatible with a pluralism about collections: a theory of objectified properties or extensions, for instance, might be better served by NF, and we have seen that such a theory might co-exist with ZFC. Indeed, this allows one to provide an appealing explanation of how the set-theoretic paradoxes arise.

Nonetheless, when it comes to the concept of set, and if set theory is to be a foundation for mathematics, then the iterative conception fares better than its currently available rivals. This vindicates the centrality of the iterative conception, and the systems it appears to sanction, in set theory and the philosophy of mathematics.

Table 8.1 *Features of conceptions of set*

Conception \ Feature	Indefinite Extensibility	Universality	Well-Foundedness	Logical	Combinatorial
Naïve	✓	✓		✓	
Iterative	✓		✓	✓	✓
Limitation of Size	✓			✓	
Definite	?	?			
Stratified		✓			
Graph	✓				✓

Bibliography

Ackermann, W. 1956, Zur Axiomatik der Mengenlehre, *Mathematische Annalen* **131**, 336–345.

Aczel, P. 1980, Frege structures and the notions of proposition, truth and set, *in* J. Barwise, J. Keisler and K. Kunen (eds), *The Kleene Symposium*, Vol. 101 of *Studies in Logic and the Foundations of Mathematics*, Elsevier, North-Holland, Amsterdam, pp. 31–59.

Aczel, P. 1988, *Non-Well-Founded Sets*, CSLI Publications, Stanford.

Anderson, A. R. and Belnap, N. D. 1975, *Entailment: The Logic of Relevance and Necessity*, Princeton University Press, Princeton, NJ.

Antonelli, A. 1999, Conceptions and paradoxes of sets, *Philosophia Mathematica* **7**, 136–163.

Awodey, S. 2004, An answer to Hellman's question: "Does category theory provide a framework for mathematical structuralism?", *Philosophia Mathematica* **12**, 54–64.

Bacon, A. 2013, Non-classical metatheory for non-classical logics, *Journal of Philosophical Logic* **42**, 335–355.

Baltag, A. 1999, STS: A structural theory of sets, *Logic Journal of the IGPL* **7**, 481–515.

Barwise, J. 1986, Situations, sets, and the axiom of foundation, *in* J. Paris, A. Wilkie and G. Wilmers (eds), *Logic Colloquium '84*, North-Holland, New York, pp. 21–36.

Barwise, J. and Etchemendy, J. 1987, *The Liar: An Essay on Truth and Circularity*, Oxford University Press, Oxford.

Barwise, J. and Moss, L. 1991, Hypersets, *Mathematical Intelligencer* **13**, 31–41.

Barwise, J. and Moss, L. 1996, *Vicious Circles*, CSLI Publications, Stanford.

Baxter, D. L. M. 1988, Identity in the loose and popular sense, *Mind* **97**, 575–582.

Beall, JC 2009, *Spandrels of Truth*, Oxford University Press, Oxford.

Beall, JC, Brady, R. T., Hazen, A. P., Priest, G. and Restall, G. 2006, Relevant restricted quantification, *Journal of Philosophical Logic* **35**, 587–598.

Bealer, G. 1982, *Quality and Concept*, Clarendon Press, Oxford.

Benacerraf, P. and Putnam, H. 1983, *Philosophy of Mathematics: Selected Readings*, 2nd edn, Cambridge University Press, Cambridge.

Bernays, P. 1935, Sur le platonisme dans les mathématiques, *L'enseignement mathématique* **34**, 52–69. English translation in Benacerraf and Putnam 1983: 258–271.

Bernays, P. 1961, Zur Frage der Unendlichkeitsschemata in der axiomatische Mengenlehre, *in* Y. Bar-Hillel, E. I. J. Poznanski, M. O. Robin and A. Robinson (eds), *Essays on the Foundations of Mathematics*, Magnes Press, Jerusalem, pp. 3–49. An English translation with a new appendix by the author appears as Bernays 1976.

Bernays, P. 1971, Zum Symposium über die Grundlagen der Mathematik, *Dialectica* **25**, 171–195. Translation by Stewe Awodey available at https://pdfs.semanticscholar.org/db74/2e06bd1951948014f8e7a9f1b9b36322ee6d.pdf.

Bernays, P. 1976, On the problem of schemata of infinity in axiomatic set theory, *in* G. Müller (ed.), *Sets and Classes: On the Work by Paul Bernays*, North-Holland, Amsterdam, pp. 121–172.

Boffa, M. 1969, Sur la théorie des ensembles sans axiome de Fondement, *Bulletin de la Société Mathématique de Belgique* **31**, 16–56.

Boolos, G. 1971, The iterative conception of set, *Journal of Philosophy* **68**, 215–231. Reprinted in Boolos 1998: 13–29.

Boolos, G. 1985, Nominalist platonism, *Philosophical Review* **94**, 327–344. Reprinted in Boolos 1998: 73–87.

Boolos, G. 1989, Iteration again, *Philosophical Topics* **17**, 5–21. Reprinted in Boolos 1998: 88–104.

Boolos, G. 1993, Whence the contradiction?, *Proceedings of the Aristotelian Society* **Supplementary Volume 67**, 213–233.

Boolos, G. 1998, *Logic, Logic, and Logic*, Harvard University Press, Cambridge, MA.

Boolos, G. 2000, Must we believe in set theory?, *in* G. Sher and R. Tieszen (eds), *Between Logic and Intuition: Essays in Honor of Charles Parsons*, Cambridge University Press, Cambridge, pp. 257–268. Reprinted in Boolos 1998: 120–132.

Booth, D. and Ziegler, R. 1996, *Finsler Set Theory: Platonism and Circularity*, Birkhäuser Verlag, Basel. Translation of Paul Finsler's papers with introductory comments.

Brady, R. T. 1984, Depth relevance of some paraconsistent logics, *Studia Logica* **43**, 63–73.

Brady, R. T. 1989, The non-triviality of dialectical set theory, *in* G. Priest, R. Routley and J. Norman (eds), *Paraconsistent Logic: Essays on the Inconsistent*, Philosophia Verlag, Munich, pp. 437–470.

Brady, R. T. 2003, Recent developments II, *in* R. Brady (ed.), *Relevant Logics and Their Rivals*, Vol. II, Ashgate, London, pp. 231–308.

Brady, R. T. 2006, *Universal Logic*, CSLI Publications, Stanford.

Brady, R. T. 2014, The simple consistency of naive set theory using metavaluations, *Journal of Philosophical Logic* **43**, 261–281.

Burgess, J. 2004, *E Pluribus Unum*: Plural logic and set theory, *Philosophia Mathematica* **3**, 193–221.

Burgess, J. 2005, *Fixing Frege*, Princeton University Press, Princeton, NJ.

Cameron, R. 2008, Turtles all the way down: Regress, priority and fundamentality, *Philosophical Quarterly* **58**, 1–14.

Cantor, G. 1883, Über unendliche, lineare Punktmannigfaltigkeiten, *Mathematische Annalen* **21**, 545–586. Reprinted in Cantor 1932: 165–209.

Cantor, G. 1887–88, Mitteilungen zur Lehre vom Transfiniten, *Zeitschrift für Philosophie und philosophische Kritik* **91/92**, 81–125 and 240–265. Reprinted in Cantor 1932: 378–439.

Cantor, G. 1895, Beiträge zur Begründung der transfiniten Mengenlehre, *Mathematische Annalen* **46**, 481–512. Reprinted in Cantor 1932: 282–311.

Cantor, G. 1899, Cantor an Dedekind, *in* Cantor 1932: 443–447. Translated in van Heijenoort 1967: 113–117.

Cantor, G. 1932, *Gesammelte Abhandlungen mathematischen und philosophischen Inhalts*, Springer, Berlin. Edited by Ernst Zermelo.

Cappelen, H. 2018, *Fixing Language: An Essay on Conceptual Engineering*, Oxford University Press, Oxford.

Chierchia, G. and Turner, R. 1988, Semantics and property theory, *Linguistics and Philosophy* **11**, 261–302.

Church, A. 1974, Set theory with a universal set, *in* L. Henkin (ed.), *Proceedings of the Tarski Symposium*, Vol. 25 of *Proceedings of Symposia in Pure Mathematics*, American Mathematical Society, Providence, RI, pp. 297–308.

Clark, P. 1993, Sets and indefinitely extensible concepts and classes, *Proceedings of the Aristotelian Society* **67**, 235–249.

Cleland, C. 1991, On the individuation of events, *Synthese* **86**, 229–254.

Cocchiarella, N. B. 1985, Frege's double correlation thesis and Quine's set theories NF and ML, *Journal of Philosophical Logic* **14**, 1–39.

Cocchiarella, N. B. 1992, Conceptual realism versus Quine on classes and higher-order logic, *Synthese* **90**, 379–436.

Coret, J. 1964, Formules stratifiées et axiome de fondation, *Comptes Rendus hebdomadaires des séances de l'Académie des Sciences de Paris série A* **264**, 809–812 and 837–839.

Cotnoir, A. 2010, Anti-symmetry and non-extensional mereology, *Philosophical Quarterly* **60**, 396–405.

Cotnoir, A. J. and Baxter, D. L. M. (eds): 2014, *Composition as Identity*, Oxford University Press, Oxford.

Crabbé, M. 1992, On NFU, *Notre Dame Journal of Formal Logic* **33**, 112–119.

Crabbé, M. 2001, Reassurance for the logic of paradox, *Review of Symbolic Logic* **4**, 479–485.

Curry, H. 1942, The inconsistency of certain formal logics, *Journal of Symbolic Logic* **7**, 115–117.

Davidson, D. 1969, The individuation of events, *in* N. Rescher (ed.), *Essays in Honor of Carl G. Hempel*, Reidel, Dordrecht, pp. 216–234. Reprinted in Davidson 2001: 123–129.

Davidson, D. 2001, *Essays on Action and Events*, 2nd edn, Clarendon Press, Oxford.

Decock, L. 2002, *Trading Ontology for Ideology: The Interplay of Logic, Set Theory and Semantics in Quine's Philosophy*, Springer, Dordrecht.

Devlin, K. 1993, *The Joy of Sets: Fundamentals of Contemporary Set Theory*, 2nd edn, Springer, New York.

Dummett, M. 1963, The philosophical significance of Gödel's theorem, *Ratio* **5**, 140–155. Reprinted in Dummett 1978: 186–201.

Dummett, M. 1978, *Truth and Other Enigmas*, Duckworth, London.

Dummett, M. 1981, *Frege: Philosophy of Language*, 2nd edn, Duckworth, London.

Dummett, M. 1991, *Frege: Philosophy of Mathematics*, Harvard University Press, Cambridge, MA.

Dummett, M. 1993, What is mathematics about?, *in* M. Dummett, *The Seas of Language*, Oxford University Press, Oxford, pp. 429–445.

Ernst, M. 2017, Category theory and foundations, *in* E. Landry (ed.), *Categories for the Working Philosopher*, Oxford University Press, Oxford, pp. 69–89.

Feferman, S. 1977, Categorical foundations and foundations of category theory, *in* R. E. Butts and J. Hintikka (eds), *Logic, Foundations of Mathematics and Computability Theory*, Vol. 1, Reidel, Dordrecht, pp. 149–165.

Feferman, S. 2004, Typical ambiguity: Trying to have your cake and eat it too, *in* G. Link (ed.), *One Hundred Years of Russell's Paradox*, Walter de Gruyter, Berlin, pp. 135–151.

Feferman, S. 2012, Foundations of unlimited category theory: What remains to be done, *Review of Symbolic Logic* **6**, 6–15.

Field, H. 1980, *Science without Numbers*, Oxford University Press, Oxford.

Field, H. 2008, *Saving Truth from Paradox*, Oxford University Press, Oxford.

Field, H. Lederman, H. and Øgaard, T. F. 2017, Prospects for a naive theory of classes, *Notre Dame Journal of Formal Logic* **58**, 461–506.

Fine, K. 1994, Essence and modality, *in* J. Tomberlin (ed.), *Philosophical Perspectives 8: Logic and Language*, Ridgeview, Atascadero, CA, pp. 1–16.

Fine, K. 1995, Ontological dependence, *Proceedings of the Aristotelian Society* **95**, 269–290.

Fine, K. 2005, Class and membership, *Journal of Philosophy* **102**, 547–572.

Finsler, P. 1926, Über die Grundlagen der Mengenlehre, I, *Mathematische Zeitschrift* **25**, 683–713. Reprinted and translated in Booth and Ziegler 1996: 103–132.

Florio, S. 2014, Unrestricted quantification, *Philosophy Compass* **9**, 441–454.

Florio, S. and Leach-Krouse, G. 2017, What Russell should have said to Burali-Forti, *Review of Symbolic Logic* **10**, 682–718.

Forster, T. 1995, *Set Theory with a Universal Set*, 2nd edn, Oxford University Press, Oxford.

Forster, T. 2003, ZF + every set is the same size as a wellfounded set, *Journal of Symbolic Logic* **68**, 1–4.

Forster, T. 2008, The iterative conception of set, *The Review of Symbolic Logic* **1**, 97–110.

Forti, M. and Honsell, F. 1983, Set theory with free construction principles, *Annali Scuola Normale Superiore di Pisa, Classe di Scienze* **10**, 493–522.

Fraenkel, A., Bar-Hillel, Y. and Lévi, A. 1973, *Foundations of Set Theory*, North-Holland, Amsterdam.

Frege, G. 1884, *Die Grundlagen der Arithmetik*, Wilhelm Koebner, Breslau. Translated as Frege 1953.

Frege, G. 1892, On concept and object, *Vierteljahrsschrift für wisseschaftliche Philosophie* **16**, 192–205. Translated in Geach and Black 1952: 42–55.

Frege, G. 1893/1903, *Grundgesetze der Arithmetik*, Vol. I and II, Verlag Hermann Pohle, Jena. Translated in Frege 2013.

Frege, G. 1953, *The Foundations of Arithmetic*, 2nd edn, Basil Blackwell, Oxford.

Frege, G. 1979, *Posthumous Writings*, Chicago University Press, Chicago, IL. Edited by Hans Hermes, Friedrich Kambartel and Friedrich Kaulbach. Translated by Peter Long and Roger White.

Frege, G. 2013, *Basic Laws of Arithmetic: Derived using Concept-Script*, Oxford University Press, Oxford. Translated by P. Ebert and M. Rossberg (with C. Wright).

Friedman, H. 1971, Higher set theory and mathematical practice, *Annals of Mathematical Logic* **2**, 325–357.

Geach, P. and Black, M. (eds): 1952, *Translations from the Philosophical Writings of Gottlob Frege*, Basic Blackwell, Oxford.

Gloede, K. 1976, Reflection principles and indescribability, *in* G. Müller (ed.), *Sets and Classes: On the Work by Paul Bernays*, North-Holland, Amsterdam, pp. 277–323.

Gödel, K. *1933, The present situation in the foundations of mathematics, *in* Gödel 1990: 45–53.

Gödel, K. 1944, Russell's mathematical logic, *in* P. A. Schilpp (ed.), *The Philosophy of Bertrand Russell*, Northwestern University, Evanston and Chicago, pp. 123–153. Reprinted in Gödel 1990: 119–141.

Gödel, K. 1947, What is Cantor's continuum problem?, *American Mathematical Monthly* **54**, 515–525. Reprinted in Gödel 1990: 176–187.

Gödel, K. *1951, Some basic theorems on the foundations of mathematics and their implications, *in* Gödel 1995: 304–323.

Gödel, K. 1958, Über eine bisher noch nicht benütze Erweitrung des finiten Standpunktes, *Dialectica* **12**, 280–287. Reprinted and translated in Gödel 1990: 240–251.

Gödel, K. 1964, What is Cantor's continuum problem?, *in* Benacerraf and Putnam 1983: 470–485. Revised and expanded version of Gödel 1947. Reprinted in Gödel 1990: 254–270.

Gödel, K. 1972, On an extension of finitary mathematics which has not yet been used, to have appeared in Dialectica. First published in Gödel 1990: 271–280. Revised and expanded English translation of Gödel 1958.

Gödel, K. 1990, *Collected Works II*, Oxford University Press, Oxford.

Gödel, K. 1995, *Collected Works III*, Oxford University Press, Oxford.

Grattan-Guiness, I. 1971, The correspondence between Georg Cantor and Philip Jourdain, *Jahresbericht der Deutschen Mathematiker-Vereinigung* **73**, 111–130.

Grišin, V. N. 1974, A nonstandard logic and its application to set theory, *Studies in Formalized Languages and Nonclassical Logics*, Nauka, Moscow, pp. 135–171. In Russian.

Hailperin, T. 1944, A set of axioms for logic, *Journal of Symbolic Logic* **9**, 1–19.

Hale, B. and Wright, C. 2001, *The Reason's Proper Study: Essays towards a Neo-Fregean Philosophy of Mathematics*, Clarendon Press, Oxford.

Hallett, M. 1984, *Cantorian Set Theory and Limitation of Size*, Clarendon Press, Oxford.

Hamacher-Hermes, A. 1994, *Inhalts- oder UmfgangsLogik?*, Alber, Freiburg.

Haslanger, S. 2000, Gender and race: What are they? What do we want them to be?, *Noûs* **34**, 31–55.

Haslanger, S. 2012, *Resisting Reality: Social Construction and Social Critique*, Oxford University Press, Oxford.

Hazen, A. 1993, Review of Pollard 1990, *Philosophia Mathematica* **1**, 173–179.

Henson, C. W. 1973, Type-raising operations on cardinal and ordinal numbers in Quine's 'New Foundations', *Journal of Symbolic Logic* **38**, 59–68.

Holmes, R. 1998, *Elementary Set Theory with a Universal Set*, Vol. 10 of *Cahiers du Centre de Logique*, Bruylant-Academia, Louvain-la-Neuve.

Holmes, R. 2001, Strong axioms of infinity in NFU, *Journal of Symbolic Logic* **66**, 87–116.

Holmes, R. 2009, Alternative axiomatic set theories, *in* E. N. Zalta (ed.), *Stanford Encyclopedia of Philosophy (Spring 2009 Edition)*. Available at http://plato.stanford .edu/archives/spr2009/entries/settheory-alternative/.

Horsten, L. 2010, Impredicative identity criteria, *Philosophy and Phenomenological Research* **80**, 411–439.

Horwich, P. 1990, *Truth*, Basil Blackwell, Oxford and Cambridge, MA.

Horwich, P. 1998, *Truth*, 2nd edn, Clarendon Press, Oxford.

Hossack, K. 2014, Sets and plural comprehension, *Journal of Philosophical Logic* **43**, 517–539.

Incurvati, L. 2008, On adopting Kripke semantics in set theory, *Review of Symbolic Logic* **1**, 81–96.

Incurvati, L. 2010, *Set Theory: Its Justification, Logic and Extent*, PhD thesis, University of Cambridge.

Incurvati, L. 2012, How to be a minimalist about sets, *Philosophical Studies* **159**, 69–87.

Incurvati, L. 2014, The graph conception of set, *Journal of Philosophical Logic* **43**, 181–208.

Incurvati, L. 2016, Can the cumulative hierarchy be categorically characterized?, *Logique et Analyse* **59**, 367–387.

Incurvati, L. 2017, Maximality principles in set theory, *Philosophia Mathematica* **25**, 159–193.

Incurvati, L. and Murzi, J. 2017, Maximally consistent sets of instances of Naive Comprehension, *Mind* **126**, 371–384.

Jané, I. 1995, The role of the absolute infinite in Cantor's conception of set, *Erkenntnis* **42**, 375–402.

Jané, I. and Uzquiano, G. 2004, Well- and non-well-founded Fregean extensions, *Journal of Philosophical Logic* **33**, 437–465.

Jech, T. 2003, *Set Theory: The Third Millenium Edition*, Springer, Berlin.

Jensen, R. 1969, On the consistency of a slight(?) modification of Quine's *New Foundations*, *Synthese* **19**, 250–263.

Johnstone, P., Power, J., Tsujishita, T., Watanabe, H. and Worrell, J. 2001, On the structure of categories of coalgebras, *Theoretical Computer Science* **260**, 87–117.

Kanamori, A. 2003, *The Higher Infinite*, 2nd edn, Springer, Berlin.

Kim, J. 1973, Causation, nomic subsumption, and the concept of event, *Journal of Philosophy* **70**, 217–236.

Kleene, S. C. 1952, *Introduction to Metamathematics*, North Holland, Amsterdam.

Koellner, P. 2003, *The Search for New Axioms*, PhD thesis, MIT, Cambridge, MA.

Koellner, P. 2009, On reflection principles, *Annals of Pure and Applied Logic* **157**, 206–219.

Kreisel, G. 1967, Informal rigour and completeness proofs, *in* I. Lakatos (ed.), *Problems in the Philosophy of Mathematics*, North-Holland, Amsterdam, pp. 138–171.

Kreisel, G. 1980, Kurt Gödel, *Biographical Memoirs of Fellows of the Royal Society of London* **26**, 149–224.

Kundera, M. 1984, *The Unbearable Lightness of Being*, Harper and Row, New York.

Ladyman, J. and Presnell, S. 2018, Does homotopy type theory provide a foundation for mathematics?, *British Journal for the Philosophy of Science* **69**, 377–420.

Lavine, S. 1994, *Understanding the Infinite*, Harvard University Press, Cambridge, MA.

Lear, J. 1977, Sets and semantics, *Journal of Philosophy* **74**, 86–102.

Leitgeb, H. 2007, On the metatheory of Field's 'Solving the paradoxes, escaping revenge', *in* JC Beall (ed.), *Revenge of the Liar*, Oxford University Press, Oxford, pp. 159–183.

Leng, M. 2007, What's there to know?, *in* M. Leng, A. Paseau and M. Potter (eds), *Mathematical Knowledge*, Oxford University Press, Oxford, pp. 84–108.

Leśniewski, S. 1916, *Podstawy ogólnej teoryi mnogości. I*, Prace Polskiego Koła Naukowego w Moskwie, Moscow. Translation in Leśniewski 1992: 129–173.

Leśniewski, S. 1992, *Foundations of the General Theory of Sets. I*, Vol. 1, Kluwer, Dordrecht.

Lévy, A. 1960, Axiom schemata of strong infinity in axiomatic set theory, *Pacific Journal of Mathematics* **10**, 223–238.

Lévy, A. and Vaught, R. 1961, Principles of partial reflection in the set theories of Zermelo and Ackermann, *Pacific Journal of Mathematics* **11**, 1045–1062.

Lewis, D. 1983, New work for a theory of universals, *Australasian Journal of Philosophy* **61**, 343–377.

Lewis, D. 1991, *Parts of Classes*, Basil Blackwell, Oxford.

Linnebo, Ø. 2008, Structuralism and the notion of dependence, *Philosophical Quarterly* **58**, 59–79.

Linnebo, Ø. 2010, Pluralities and sets, *Journal of Philosophy* **107**, 144–164.

Linnebo, Ø. 2018, Dummett on indefinite extensibility, *Philosophical Issues* **28**, 196–220.

Linnebo, Ø. and Pettigrew, R. 2011, Category theory as an autonomous foundation, *Philosophia Mathematica* **19**, 227–254.

Linnebo, Ø. and Rayo, A. 2012, Hierarchies ontological and ideological, *Mind* **121**, 269–308.

Lipton, P. 2003, *Inference to the Best Explanation*, Routledge, London and New York.

Lowe, E. J. 1989, What is a criterion of identity?, *Philosophical Quarterly* **39**, 1–21.

Lowe, E. J. 1991, One-level versus two-level identity criteria, *Analysis* **51**, 192–194.

Lowe, E. J. 2003, Individuation, *in* M. Loux and D. Zimmerman (eds), *Oxford Handbook of Metaphysics*, Oxford University Press, Oxford, pp. 75–95.

Lowe, E. J. 2005, Ontological dependence, *in* E. N. Zalta (ed.), *Stanford Encyclopedia of Philosophy (Fall 2008 Edition)*. Available at http://plato.stanford.edu/archives/fall2008/entries/dependence-ontological/.

Lowe, E. J. 2007, Sortals and the individuation of objects, *Mind and Language* **22**, 514–533.

MacLane, S. 1986, *Mathematics, Form and Function*, Springer-Verlag, New York.

Maddy, P. 1983, Proper classes, *Journal of Symbolic Logic* **48**, 113–139.

Maddy, P. 1988, Believing the axioms. I, *Journal of Symbolic Logic* **53**, 481–511.

Maddy, P. 1990, *Realism in Mathematics*, Clarendon Press, Oxford.

Maddy, P. 1996, Set theoretic naturalism, *Journal of Symbolic Logic* **61**, 490–514.

Maddy, P. 1997, *Naturalism in Mathematics*, Clarendon Press, Oxford.

Maddy, P. 2007, *Second Philosophy: A Naturalistic Method*, Oxford University Press, New York.

Maddy, P. 2011, *Defending the Axioms: On the Philosophical Foundations of Set Theory*, Oxford University Press, New York.

Martin, D. 1970, Review of *Set Theory and Its Logic* by Willard Van Orman Quine, *Journal of Philosophy* **67**, 111–114.

Martin, D. A. 1975, Borel determinacy, *Annals of Mathematics* **102**, 363–371.

Martin, D. A. 1998, Mathematical evidence, *in* H. G. Dales and G. Oliveri (eds), *Truth in Mathematics*, Clarendon Press, Oxford, pp. 215–230.

Martin, D. A. 2001, Multiple universes of sets and indeterminate truth-values, *Topoi* **20**, 5–16.

Martin, D. A. and Steel, J. R. 1989, A proof of projective determinacy, *Journal of the American Mathematical Society* **2**, 71–125.

Mathias, A. 2001, Slim models of Zermelo set theory, *Journal of Symbolic Logic* **66**, 487–496.

McGee, V. 1992, Two problems with Tarski's theory of consequence, *Proceedings of the Aristotelian Society* **92**, 273–292.

McGee, V. 1997, How we learn mathematical language, *Philosophical Review* **106**, 35–68.

McLarty, C. 1992, Failure of cartesian closedness in NF, *Journal of Symbolic Logic* **57**, 555–556.

McLarty, C. 1993, Anti-foundation and self-reference, *Journal of Philosophical Logic* **22**, 19–28.

Meadows, T. 2015, Unpicking Priest's bootstraps, *Thought* **4**, 181–188.

Menzel, C. 2014, Wide sets, ZFCU, and the iterative conception, *Journal of Philosophy* **111**, 57–83.

Meschkowski, H. 1967, *Probleme des Unendlichen: Werk und Leben Georg Cantors*, Vieweg, Braunschweig.

Meyer, R., Routley, R. and Dunn, J. M. 1979, Curry's Paradox, *Analysis* **39**, 124–128.

Mirimanoff, D. 1917, Remarque sur la théorie des ensembles et les antinomies Cantorienne. I, *L'Enseignment Mathématique* **19**, 37–52.

Mitchell, E. 1976, *A Model of Set Theory with a Universal Set*, PhD thesis, University of Wisconsin at Madison.

Montague, R. 1957, *Contributions to the Axiomatic Foundation of Set Theory*, PhD thesis, University of California, Berkeley.

Montague, R. 1961, Fraenkel's addition to the axioms of Zermelo, *in* Y. Bar-Hillel, E. I. J. Poznanski, M. O. Rabin and A. Robinson (eds), *Essays on the Foundations of Mathematics*, Magnes Press, Jerusalem, pp. 91–114.

Morris, S. 2017, The significance of Quine's New Foundations for the philosophy of set theory, *The Monist* **100**, 167–179.

Moschovakis, Y. N. 2006, *Notes on Set Theory*, 2nd edn, Springer, New York.

Moss, L. 2009, Non-wellfounded set theory, *in* E. N. Zalta (ed.), *Stanford Encyclopedia of Philosophy (Fall 2009 Edition)*. Available at http://plato.stanford.edu/archives/fall2009/entries/nonwellfounded-set-theory/.

Myhill, J. 1984, Paradoxes, *Synthese* **60**, 129–143.

Oliver, A. 2000, Logic, mathematics and philosophy: Review of Boolos 1998, *British Journal for the Philosophy of Science* **51**, 857–873.

Oliver, A. and Smiley, T. 2006, What are sets and what are they for?, *Philosophical Perspectives* **20**, 123–155.

Oliver, A. and Smiley, T. 2012, *Plural Logic*, Oxford University Press, Oxford.

Orey, S. 1964, New Foundations and the Axiom of Counting, *Notre Dame Journal of Formal Logic* **31**, 655–660.

Parikh, R. 1971, Existence and feasibility in arithmetic, *Journal of Symbolic Logic* **36**, 494–508.

Parsons, C. 1974, Sets and classes, *Noûs* **8**, 1–12.

Parsons, C. 1977, What is the iterative conception of set?, *in* R. E. Butts and J. Hintikka (eds), *Logic, Foundations of Mathematics, and Computability Theory. Proceedings of the Fifth International Congress of Logic, Methodology and the Philosophy of Science (London, Ontario, 1975)*, Reidel, Dordrecht. Reprinted in Benacerraf and Putnam 1983: 503–529.

Parsons, C. 2008, *Mathematical Thought and Its Objects*, Cambridge University Press, Cambridge.

Parsons, C. 2014, Analyticity for realists, *in* J. Kennedy (ed.), *Interpreting Gödel*, Cambridge University Press, Cambridge, pp. 131–150.

Paseau, A. 2007, Boolos on the justification of set theory, *Philosophia Mathematica* **15**, 30–53.

Pollard, S. 1988, Plural quantification and the axiom of choice, *Philosophical Studies* **54**, 393–397.

Pollard, S. 1990, *Philosophical Introduction to Set Theory*, Notre Dame University Press, Notre Dame.

Pollard, S. 1992, Choice again, *Philosophical Studies* **66**, 285–296.

Potter, M. 2004, *Set Theory and Its Philosophy*, Oxford University Press, Oxford.

Potter, M. 2009, Abstractionist class theory: is there any such thing?, *in* J. Lear and A. Oliver (eds), *The Force of Argument: Essays in Honor of Timothy Smiley*, Routledge, London, pp. 186–204.

Priest, G. 1979, The logic of paradox, *Journal of Philosophical Logic* **8**, 219–241.

Priest, G. 2001, Minimally inconsistent LP, *Studia Logica* **50**, 321–331.

Priest, G. 2002, *Beyond the Limits of Thought*, 2nd, extended edn, Oxford University Press, Oxford.

Priest, G. 2006a, *Doubt Truth to Be a Liar*, Oxford University Press, New York.

Priest, G. 2006b, *In Contradiction*, 2nd edn, Oxford University Press, Oxford.

Priest, G. 2017, What *If*? the exploration of an idea, *Australasian Journal of Logic* **14**, 54–127.

Priest, G. and Routley, R. 1989, Applications of paraconsistent logic, *in* G. Priest, R. Routley and J. Norman (eds), *Paraconsistent Logic: Essays on the Inconsistent*, Philosophia Verlag, Hamden, pp. 369–373.

Quine, W. V. 1937, New foundations for mathematical logic, *American Mathematical Monthly* **44**, 70–80.

Quine, W. V. 1938a, On Cantor's Theorem, *Journal of Symbolic Logic* **2**, 120–124.

Quine, W. V. 1938b, On the theory of types, *Journal of Symbolic Logic* **3**, 125–139.

Quine, W. V. 1940, *Mathematical Logic*, Norton, New York.

Quine, W. V. 1941, Whitehead and the rise of modern logic, *in* P. A. Schilpp (ed.), *The Philosophy of Alfred North Whitehead*, Northwestern University, Evanston and Chicago, pp. 127–163. Reprinted in Quine 1995: 3–36.

Quine, W. V. 1951a, *Mathematical Logic*, revised edition of Quine 1940 edn, Harvard University Press, Cambridge, MA.

Quine, W. V. 1951b, Two dogmas of empiricism, *Philosophical Review* **60**, 20–43.

Quine, W. V. 1956, Unification of universes in set theory, *Journal of Symbolic Logic* **21**, 267–279.

Quine, W. V. 1963, *Set Theory and Its Logic*, Harvard University Press, Cambridge, MA.

Quine, W. V. 1970, Reply to D. A. Martin, *Journal of Philosophy* **67**, 247–248.

Quine, W. V. 1987a, The inception of New Foundations. In Quine 1995: 286–289.

Quine, W. V. 1987b, *Quiddities: An Intermittently Philosophical Dictionary*, Harvard University Press, Cambridge, MA.

Quine, W. V. 1995, *Selected Logic Papers*, Harvard University Press, Cambridge, MA.

Ramsey, F. P. 1925, The foundations of mathematics, *Proceedings of the London Mathematical Society* **25**, 338–384. Reprinted in Ramsey 1990: 164–224.

Ramsey, F. P. 1990, *Philosophical Papers*, Cambridge University Press, Cambridge. Edited by D. H. Mellor.

Rawls, J. 1971, *A Theory of Justice*, Harvard University Press, Cambridge, MA.

Rayo, A. and Uzquiano, G. 1999, Toward a theory of second-order consequence, *Notre Dame Journal of Formal Logic* **40**, 315–325.

Rayo, A. and Williamson, T. 2003, A completeness theorem for unrestricted first-order languages, *in* JC Beall (ed.), *Liar and Heaps*, Clarendon Press, Oxford, pp. 331–356.

Reinhardt, W. N. 1974, Remarks on reflection principles, large cardinals, and elementary embeddings, *in* T. J. Jech (ed.), *Axiomatic Set Theory II: Proceedings of Symposia in Pure Mathematics, 13*, American Mathematical Society, Providence, RI, pp. 189–205.

Restall, G. 1992, A note on naive set theory in LP, *Notre Dame Journal of Formal Logic* **33**, 422–432.

Rieger, A. 2000, An argument for Finsler-Aczel set theory, *Mind* **109**, 241–253.

Ripley, D. 2015, Naive set theory and nontransitive logic, *Review of Symbolic Logic* pp. 553–571.

Rosser, J. B. 1942, The Burali-Forti Paradox, *Journal of Symbolic Logic* **7**, 1–17.

Rosser, J. B. 1953, *Logic for Mathematicians*, McGraw-Hill, New York.

Rosser, J. B. and Wang, H. 1950, Non-standard models for formal logics, *Journal of Symbolic Logic* **15**, 113–129.

Routley, R. 1980, *Exploring Meinong's Jungle and Beyond: An Investigation of Noneism and the Theory of Items*, Philosophy Department, RSSS, Australian National University, Canberra.

Ruffino, M. 2003, Why Frege would not be a neo-Fregean, *Mind* **112**, 51–78.

Russell, B. 1902, Letter to Frege. In van Heijenoort 1967: 124–125.

Russell, B. 1903, *Principles of Mathematics*, Allen & Unwin, London.

Russell, B. 1906, On some difficulties in the theory of transfinite numbers and order types, *Proceedings of the London Mathematical Society* **4**, 29–53.

Russell, B. 1959, *My Philosophical Development*, Routledge, London.

Rutten, J. 2000, Universal coalgebra: A theory of systems, *Theoretical Computer Science* **249**, 3–80.

Schaffer, J. 2009, On what grounds what, *in* D. Chalmers, D. Manley and R. Wasserman (eds), *Metametaphysics*, Oxford University Press, Oxford, pp. 347–383.

Scharp, K. 2013, *Replacing Truth*, Oxford University Press, Oxford.

Schindler, T. 2019, Classes, why and how, *Philosophical Studies* **176**, 407–435.

Scott, D. 1955, Definitions by abstraction in axiomatic set theory, *Bulletin of the American Mathematical Society* **61**, 442.

Scott, D. 1960, A different kind of model for set theory. Unpublished paper given at the Stanford Congress of Logic, Methodology and Philosophy of Science.

Scott, D. 1961, Measurable cardinals and constructible sets, *Bulletin de l'Académie Polonaise des Sciences* **9**, 521–524.

Scott, D. 1974, Axiomatizing set theory, *in* T. J. Jech (ed.), *Axiomatic Set Theory II. Proceedings of Symposia in Pure Mathematics, 13*, American Mathematical Society, Providence, RI, pp. 207–214.

Shapiro, S. 1991, *Foundations without Foundationalism: A Case for Second-Order Logic*, Oxford University Press, Oxford.

Shapiro, S. 1997, *Philosophy of Mathematics: Structure and Ontology*, Oxford University Press, New York.

Shapiro, S. 2000, *Thinking about Mathematics*, Oxford University Press, New York.

Shapiro, S. 2003, Prolegomenon to any future neo-logicist set theory: Abstraction and indefinite extensibility, *British Journal for the Philosophy of Science* **54**, 59–91.

Shapiro, S. 2004, Foundations of mathematics: Metaphysics, epistemology, structure, *Philosophical Quarterly* **54**, 16–37.

Shapiro, S. 2006a, Computability, proof, and open-texture, *in* A. Olszweski, J. Woleński and R. Janusz (eds), *Church's Thesis After 70 Years*, Ontos-Verlag, Heusenstamm, pp. 420–451.

Shapiro, S. 2006b, Structure and identity, *in* F. MacBride (ed.), *Identity and Modality*, Oxford University Press, Oxford, pp. 109–145.

Shapiro, S. 2008, Identity, indiscernibility, and *ante rem* structuralism: The tale of i and $-i$, *Philosophia Mathematica* **16**, 1–26.

Shapiro, S. and Weir, A. 1999, New V, ZF and abstraction, *Philosophia Mathematica* **7**, 293–321.

Shapiro, S. and Wright, C. 2006, All things indefinitely extensible, *in* A. Rayo and G. Uzquiano (eds), *Absolute Generality*, Oxford University Press, Oxford, pp. 255–304.

Shoenfield, J. R. 1965, *Mathematical Logic*, Addison-Wesley, Reading, MA.

Shoenfield, J. R. 1977, Axioms of set theory, *in* J. Barwise (ed.), *Handbook of Mathematical Logic*, North-Holland, Amsterdam, pp. 321–344.

Sider, T. 2007, Parthood, *Philosophical Review* **116**, 51–91.

Simons, P. 1991, Part/whole ii: Mereology since 1900, *in* H. Burkhardt and B. Smith (eds), *Handbook of Metaphysics and Ontology*, Philosophia, Munich, pp. 209–210.

Simpson, S. G. 1999, *Subsystems of Second-Order Arithmetic*, Springer-Verlag, Berlin.

Soames, S. 1999, *Understanding Truth*, Oxford University Press, Oxford.

Specker, E. P. 1953a, The Axiom of Choice in Quine's New Foundations for Mathematical Logic, *Proceedings of the National Academy of Sciences of the United States of America* **39**, 972–975.

Specker, E. P. 1953b, Dualtität, *Dialectica* **12**, 451–465.

Specker, E. P. 1962, Typical ambiguity, *in* E. Nagel (ed.), *Logic, Methodology and Philosophy of Science*, Stanford University Press, Stanford, CA, pp. 116–123.

Strawson, P. 1979, *Individuals*, University Paperbacks, London.

Sullivan, P. 2007, How did Frege fall into the contradiction?, *Ratio* **20**, 91–107.

Tait, W. W. 1998, Zermelo's conception of set theory and reflection principles, *in* M. Schirn (ed.), *The Philosophy of Mathematics Today*, Oxford University Press, Oxford, pp. 469–483.

Tait, W. W. 2001, Gödel's unpublished papers in the foundations of mathematics, *Philosophia Mathematica* **9**, 87–126.

Tarski, A. 1955, The notion of rank in axiomatic set theory and some of its applications, *Bulletin of the American Mathematical Society* **61**, 443.

Thomas, M. 2014, Expressive limitations of naïve set theory in LP and minimally inconsistent LP, *Review of Symbolic Logic* **7**, 341–350.

Thomas, M. 2018, Approximating cartesian closed categories in NF-style set theories, *Journal of Philosophical Logic* **47**, 143–160.

Trueman, R. 2015, The concept *horse* with no name, *Philosophical Studies* **172**, 1889–1906.

Turi, D. and Rutten, J. 1998, On the foundations of final coalgebra semantics: non-well-founded sets, partial orders, metric spaces, *Mathematical Structures in Computer Science* **8**, 481–540.

Turing, A. 1936, On computable numbers, with an application to the *Entscheidungsproblem*, *Proceedings of the London Mathematical Society* **42**, 230–265.

Tutte, W. T. 2001, *Graph Theory*, paperback edn, Cambridge University Press, Cambridge.

Urquhart, A. 1988, Russell's zigzag path to the ramified theory of types, *Russell: The Journal of Bertrand Russell Studies* **8**, 82–91.

Uzquiano, G. 1999, Models of second-order Zermelo set theory, *Bulletin of Symbolic Logic* **5**, 289–302.

Uzquiano, G. 2015a, Modality and paradox, *Philosophy Compass* **10**, 284–300.

Uzquiano, G. 2015b, A neglected resolution of Russell's paradox of propositions, *Review of Symbolic Logic* **8**, 328–344.

van den Berg, B. and De Marchi, F. 2007, Non-well-founded trees in categories, *Annals of Pure and Applied Logic* **146**, 40–59.

van Heijenoort, J. 1967, *From Frege to Gödel: A Source Book in Mathematical Logic, 1879–1931*, Harvard University Press, Cambridge, MA.

van Inwagen, P. 1990, *Material Beings*, Cornell University Press, Ithaca, NY.

Varzi, A. 2008, The extensionality of parthood and composition, *Philosophical Quarterly* **58**, 108–133.

von Neumann, J. 1923, Letter to Ernst Zermelo, 15 August 1923. In Meschkowski 1967: 271–273.

von Neumann, J. 1925, Eine Axiomatisierung der Mengenlehre, *Journal für die reine und angewandte Mathematik* **154**, 219–240. English translation in van Heijenoort 1967: 393–413.

Waismann, F. 1945, Verifiability, *Proceedings of the Aristotelian Society* **Supplementary Volume 19**, 119–150.

Wang, H. 1950, A formal system of logic, *Journal of Symbolic Logic* **15**, 25–32.

Wang, H. 1974, *From Mathematics to Philosophy*, Routledge and Kegan Paul, London.

Wang, H. 1977, Large sets, *in* R. E. Butts and J. Hintikka (eds), *Logic, Foundations of Mathematics, and Computability Theory: Proceedings of the Fifth International Congress of Logic, Methodology and the Philosophy of Science (London, Ontario, 1975)*, Reidel, Dordrecht, pp. 309–334.

Wang, H. 1996, *A Logical Journey: From Gödel to Philosophy*, MIT Press, Cambridge, MA.

Weber, Z. 2010a, Extensionality and restriction in naive set theory, *Studia Logica* **94**, 87–104.

Weber, Z. 2010b, Transfinite numbers in paraconsistent set theory, *Review of Symbolic Logic* **3**, 71–92.

Weber, Z. 2012, Transfinite cardinals in paraconsistent set theory, *Review of Symbolic Logic* **5**, 269–292.

Weber, Z. 2013, Notes on inconsistent set theory, *in* K. Tanaka, F. Berto, E. Mares and F. Paoli (eds), *Paraconsistency: Logic and Applications*, Springer, Dordrecht, pp. 315–328.

Weir, A. 1998a, Naïve set theory is innocent!, *Mind* **107**, 763–798.

Weir, A. 1998b, Naïve set theory, paraconsistency and indeterminacy: part I, *Logique et Analyse* **161–163**, 219–266.

Weir, A. 1999, Naïve set theory, paraconsistency and indeterminacy: part II, *Logique et Analyse* **167–168**, 283–340.

Weir, A. 2004, There are no true contradictions, *in* G. Priest, JC Beall and B. Armour-Garb (eds), *The Law of Non-Contradiction: New Philosophical Essays*, Clarendon Press, Oxford, pp. 385–417.

Welch, P. 2014, Global reflection principles, *in* H. Leitgeb, I. Niiniluoto, P. Seppälä and E. Sober (eds), *Logic, Methodology and Philosophy of Science: Proceedings of the Fifteenth International Congress*, College Publications, London, pp. 82–100.

Welch, P. and Horsten, L. 2016, Reflecting on absolute infinity, *Journal of Philosophy* **113**, 89–111.

Whitehead, A. N. and Russell, B. 1910, 1912, 1913, *Principa Mathematica*, Cambridge University Press, Cambridge. 3 volumes.

Williamson, T. 1990, *Identity and Discrimination*, Basil Blackwell, Oxford.

Williamson, T. 1991, Fregean directions, *Analysis* **51**, 194–195.

Williamson, T. 1994, Never say never, *Topoi* **13**, 135–145.

Williamson, T. 2007, *The Philosophy of Philosophy*, Blackwell, Oxford.

Woods, J. 2003, *Paradox and Paraconsistency: Conflict Resolution in the Abstract Sciences*, Cambridge University Press, Cambridge.

Wright, C. 1983, *Frege's Conception of Numbers as Objects*, Aberdeen University Press, Aberdeen.

Zermelo, E. 1904, Beweis daß jede Menge wohlgeordnet werden kann, *Mathematische Annalen* **59**, 514–516.

Zermelo, E. 1908, Untersuchungen über die Grundlagen der Mengenlehre I, *Mathematische Annalen* **65**, 261–281. Reprinted with translation in Zermelo 2010: 188–229.

Zermelo, E. 1930, Über Grenzzahlen und Mengenbereiche: Neue Untersuchungen über die Grundlagen der Mengenlehre, *Fundamenta Mathematicae* **16**, 29–47. Reprinted and translated in Zermelo 2010: 400–431.

Zermelo, E. 2010, *Collected Works*, Vol. I, Springer-Verlag, Berlin. Edited by Heinz-Dieter Ebbinghaus, Craig G. Fraser and Akihiro Kanamori.

Index

Printed in the United States
by Baker & Taylor Publisher Services